METHODS IN MOLECULAR BIOLOGY

Series Editor
John M. Walker
School of Life and Medical Sciences
University of Hertfordshire
Hatfield, Hertfordshire, AL10 9AB, UK

For further volumes:
http://www.springer.com/series/7651

Celiac Disease

Methods and Protocols

Edited by

Anthony W. Ryan

Department of Clinical Medicine, Trinity College Dublin; Institute of Molecular Medicine, Trinity Centre for Health Sciences, St James's Hospital, Dublin, Ireland.

Editor
Anthony W. Ryan
Department of Clinical Medicine
Trinity College Dublin
Dublin, Ireland

Institute of Molecular Medicine
Trinity Centre for Health Sciences
St James's Hospital
Dublin, Ireland

ISSN 1064-3745   ISSN 1940-6029   (electronic)
Methods in Molecular Biology
ISBN 978-1-4939-2838-5   ISBN 978-1-4939-2839-2   (eBook)
DOI 10.1007/978-1-4939-2839-2

Library of Congress Control Number: 2015944312

Springer New York Heidelberg Dordrecht London

Printed on acid-free paper

Humana Press is a brand of Springer
Springer Science+Business Media LLC New York is part of Springer Science+Business Media (www.springer.com)

Preface

Recent decades have seen considerable advances in our understanding of celiac disease. The condition, once thought to be limited to individuals of European ancestry, has been discovered at varying prevalence in North Africa, the Middle East, India, and China. The precipitating auto-antigen has been characterized. The genetic association of the HLA region was discovered early and refined in the years that followed. However, conclusive identification of non-HLA genetic risk proved elusive until the advent of genome-wide association studies, which have extended our understanding of the genetic component far beyond what could have been envisaged a short time ago. Current estimates suggest that more than 50% of the population variability associated with celiac disease risk can be explained by known genetic loci. Despite these advances, there is still a great deal to be learned, both about the nature of genetic risk and the functional genomic consequences of the established risk factors.

Building on this knowledge will require detailed molecular analysis of the associated pathways and many cell types involved in the disease, as well as embracing new technologies such as next-generation sequencing. At the same time, long established molecular and immunology methods will continue to have a place for some time to come. This book brings together novel and more traditional methods in molecular biology and immunology, in order to provide a tool-kit for all stages of celiac disease research, from the practicalities of obtaining high-quality samples, to molecular analysis and bioinformatics.

Part I of this book sets the background with a number of reviews to describe the history and nature of the disease, its diagnosis, the role of animal models, and study designs for investigating genetic susceptibility. Part II describes the molecular techniques, including tissue culture, isolation and cloning of relevant cell types, high content analysis of biopsies, and HLA genotyping. Subsequent chapters describe analyses of gene expression and functional analysis of genetic variants: detecting allelic expression imbalance, reporter assays, and siRNA knockdown. The final chapter in this part describes a number of protocols for epigenetic analysis, which has attracted little attention until recently.

A great deal of modern molecular biology relies on high-throughput automated analyses, which produce vast quantities of data. Coming to grips with this flow of information represents a new set of challenges, dealt with in Part III. The final three chapters describe analysis pipelines for bioinformatic prediction of antigenicity, quality control and analysis of GWAS data, and transcriptome analysis by next-generation sequencing. These techniques require a certain level of scripting skills, not commonly held by laboratory based researchers. For this reason, Part III begins with an outline of scripting for data management, focusing on tools which are freely available to researchers who wish to explore them.

To conclude, I would like to thank John Walker and Humana Press for their guidance and assistance throughout this project, and I extend my gratitude to all contributors to the book, who took time out from their busy research and clinical schedules to write their chapters.

Dublin, Ireland   *Anthony W. Ryan*

Contents

Contributors

RICHARD J.L. ANNEY • *Institute of Molecular Medicine, Trinity Centre for Health Sciences, St James's Hospital, Dublin, Ireland; Department of Psychiatry, Trinity College Dublin, Dublin, Ireland*

RENATA AURICCHIO • *Department of Translational Medicine, Section of Pediatrics, University of Naples Federico II, Via S. Pansini 5, Naples, Italy; European Laboratory for the Investigation of Food Induced Diseases (ELFID), University of Naples Federico II, Via S. Pansini 5, Naples, Italy*

ANNE-MARIE BAIRD • *Department of Clinical Medicine, Trinity College Dublin, Dublin, Ireland; Thoracic Oncology Research Group, Institute of Molecular Medicine, Trinity Centre for Health Science, St. James's Hospital, Dublin, Ireland; Cancer and Aging Research Program, Queensland University of Technology, Brisbane, Australia*

GREG BYRNE • *School of Biological Sciences, Dublin Institute of Technology, Dublin, Ireland*

MIKAELA M. BYRNE • *Department of Clinical Medicine, Trinity College Dublin, Dublin, Ireland; Institute of Molecular Medicine, Trinity College Dublin, Dublin, Ireland*

CIARA COLEMAN • *Department of Medicine, Institute of Molecular Medicine, Trinity College Dublin, College Green, Dublin, Ireland; St. James's Hospital, Dublin, Ireland*

SARAH E.J. COOPER • *Immunology Department, Trinity College Dublin, Dublin, Ireland*

SANDRA D' ALFONSO • *Department of Health Sciences, University of Eastern Piedmont, Novara, Italy; Interdisciplinary Research Centre of Autoimmune Disease (IRCAD), University of Eastern Piedmont, Novara, Italy*

ENNIA DAMETTO • *Immunogenetica e Biologia dei Trapianti, AOU Città della salute e della scienza di torino, via Santena, Torino, Italy*

ANTHONY MITCHELL DAVIES • *INCHSA, Institute for Molecular Medicine, Trinity College Dublin, Dublin, Ireland*

JEAN DUNNE • *Immunology Department, Trinity College Dublin, Dublin, Ireland; Immunology Department, St James's Hospital, Dublin, Ireland*

MARGARET R. DUNNE • *Department of Surgery, Trinity Centre for Health Sciences, St James's Hospital, Dublin, Ireland; National Children's Research Centre, Our Lady's Children's Hospital, Dublin, Ireland; Department of Immunology, Institute of Molecular Medicine, St. James's Hospital, Trinity College Dublin, Dublin, Ireland*

LOUISE ELLIOT • *Immunology Department, Trinity College Dublin, Dublin, Ireland*

MARIA EDVIGE FASANO • *Immunogenetica e Biologia dei Trapianti, AOU Città della salute e della scienza di torino, via Santena, Torino, Italy*

CONLETH F. FEIGHERY • *Immunology Department, Trinity College Dublin, Dublin, Ireland; Immunology Department, St. James's Hospital, Dublin, Ireland*

MICHAEL FREELEY • *Department of Medicine, Institute of Molecular Medicine, Trinity College Dublin, St. James's Hospital, Dublin, Ireland*

JILLIAN M. GAHAN • *Department of Clinical Medicine, Trinity College Dublin, Dublin, Ireland; Institute of Molecular Medicine, Trinity Centre for Health Sciences, St James's Hospital, Dublin, Ireland*

MARTINA GALATOLA • *Department of Translational Medicine, Section of Pediatrics, University of Naples Federico II, Via S. Pansini 5, Naples, Italy; European Laboratory for the Investigation of Food Induced Diseases (ELFID), University of Naples Frederico II, Via S. Pansini 5, Naples, Italy*

STEVEN G. GRAY • *Department of Clinical Medicine, Trinity College Dublin, Dublin, Ireland; Thoracic Oncology Research Group, Institute of Molecular Medicine, Trinity Centre for Health Science, St. James's Hospital, Dublin, Ireland; HOPE Directorate, St James's Hospital, Dublin, Ireland*

LUIGI GRECO • *Department of Translational Medicine, Section of Pediatrics, University of Naples Federico II, Via S. Pansini 5, Naples, Italy; European Laboratory for the Investigation of Food Induced Diseases (ELFID), University of Naples Federico II, Naples, Italy*

MATTHEW HILL • *Neurosciences and Mental Health Research Institute, Cardiff University School of Medicine, Hadyn Ellis Building, Cathays, Cardiff, UK*

KARSTEN HOKAMP • *School of Genetics and Microbiology, Smurfit Institute of Genetics, Trinity College Dublin, College Green, Dublin, Ireland*

JOSEPHINE M. JU • *Department of Immunology, Mayo Clinic Rochester, Rochester, MN, USA*

JACINTA KELLY • *National Children's Research Centre, Our Lady's Children's Hospital, Dublin, Ireland*

FRITS KONING • *Department of Immunohematology and Blood Transfusion, Leiden University Medical Centre, Leiden, The Netherlands*

YVONNE KOOY-WINKELAAR • *Department of Immunohematology and Blood Transfusion, Leiden University Medical Centre, Leiden, The Netherlands*

AIDEEN LONG • *Department of Medicine, Institute of Molecular Medicine, Trinity College Dublin, St. James's Hospital, Dublin, Ireland*

ERIC V. MARIETTA • *Department of Immunology, Mayo Clinic Rochester, Rochester, MN, USA; Department of Gastroenterology, Mayo Clinic Rochester, Rochester, MN, USA; Department of Dermatology, Mayo Clinic Rochester, Rochester, MN, USA*

ROSS MCMANUS • *Department of Medicine, Institute of Molecular Medicine, Trinity College Dublin, Dublin, Ireland*

BASHIR M. MOHAMED • *Immunology Department, Trinity College Dublin, Dublin, Ireland; Thoracic Oncology, St James's Hospital, Dublin, Ireland*

BEN MOLLOY • *Department of Medicine, Institute of Molecular Medicine, Trinity College Dublin, Dublin, Ireland*

ROSS T. MURPHY • *Department of Cardiology, St James's Hospital, Dublin, Ireland; Institute of Cardiovascular Science, St. James's Hospital, Dublin, Ireland*

JOSEPH A. MURRAY • *Department of Immunology, Mayo Clinic Rochester, Rochester, MN, USA; Department of Gastroenterology, Mayo Clinic Rochester, Rochester, MN, USA; Department of Gastroenterology and Hepatology, Mayo Foundation, Rochester, MN, USA*

ÅSA TORINSSON NALUAI • *Institute of Biomedicine, Sahlgrenska Academy at the University of Gothenburg, Gothenburg, Sweden*

CATHAL P. O'BRIEN • *St James's Hospital, Dublin, Ireland; Trinity College Dublin, Dublin, Ireland*

ANTOINETTE S. PERRY • *Prostate Molecular Oncology, Institute of Molecular Medicine, Trinity College Dublin, Dublin, Ireland*

EMMA M. QUINN • *Department of Clinical Medicine, Institute of Molecular Medicine, Trinity College Dublin, Dublin, Ireland*

ANTHONY W. RYAN • *Department of Clinical Medicine, Trinity College Dublin, Dublin, Ireland; Institute of Molecular Medicine, Trinity Centre for Health Sciences, St James's Hospital, Dublin, Ireland*

GRAHAM D. TURNER • *Department of Clinical Medicine, Trinity College Dublin, Dublin, Ireland; Institute of Molecular Medicine, Trinity Centre for Health Sciences, St. James's Hospital, Dublin, Ireland*

SHARON WILSON • *Institute of Molecular Medicine, Trinity College Dublin, Dublin, Ireland*

Part I

Background

Chapter 1

Celiac Disease: Background and Historical Context

Graham D. Turner, Margaret R. Dunne, and Anthony W. Ryan

Abstract

Medical descriptions of celiac disease date to the first century BC, and the first modern description was published in 1888. Further insights were gained throughout the 1900s, culminating in the identification of the dietary component, the major genetic determinant, and the autoantigen by the turn of the century. Understanding of the age of onset, population prevalence, and the extent of subclinical celiac disease developed in tandem. Thanks to advances in genomics, currently established loci account for over 50 % of the genetic risk. Nonetheless, much remains to be discovered. Advances in high-throughput genomic, biochemical, and cell analyses, as well as the bioinformatics needed to process the data, promise to deepen our understanding further. Here we present a primer of celiac disease, viewing the condition in turn from the historical, epidemiological, immunological, molecular, and genetic points of view. Research into any ailment has specific requirements: study subjects must be identified and relevant tissue samples collected and stored with the appropriate timing and conditions. These requirements are summarized. To conclude, a short discussion of future prospects is presented.

Key words Celiac disease, Genetic risk, HLA, Gluten, Transglutaminase, T cell

1 A Brief Description of Celiac Disease

Celiac disease (CD) is an autoimmune inflammatory condition which primarily affects the small intestine. Common symptoms include bloating, diarrhea, and, particularly in children, "failure to thrive" due to malabsorption of dietary nutrients [1]. In a subset of cases the disease also manifests as dermatitis herpetiformis, an inflammatory, blistering condition of the skin [2]. At the molecular level, CD is triggered by an immune reaction to gluten, a protein present in wheat, barley, and rye. Virtually all sufferers of CD carry at least one copy of either HLA-DQ2 or HLA-DQ8, two genetic variants of the human leukocyte antigen (HLA, also termed major histocompatibility complex, MHC) Class II molecules [3]. However, these variants are common in many populations [4], while only about 1 % develop CD. As such, possession of one or both of the variants is necessary, but not sufficient, for onset of the disease.

Anthony W. Ryan (ed.), *Celiac Disease: Methods and Protocols*, Methods in Molecular Biology, vol. 1326,
DOI 10.1007/978-1-4939-2839-2_1, © Springer Science+Business Media New York 2015

Other factors, both genetic and environmental, come into play in its development.

The gold standard for diagnosis of CD requires endoscopic intestinal biopsy and serological detection of anti-tissue transglutaminase (anti-tTG) and/or anti-endomysial antibodies (anti-EMA). At the histological level, CD leads to flattened microvilli in the small intestine of sufferers. This villous atrophy, which may be graded according to the Marsh classification [5], is accompanied by an increase in the number of intraepithelial lymphocytes, indicative of the immune component of the disease. CD may be accompanied by additional comorbidities, including osteoporosis, other autoimmune diseases, and psychiatric disorders [6]. The characteristic villous atrophy explains the malabsorption associated with the disease, which may also lead to secondary lactose intolerance, as the cells that normally produce the lactase enzyme are damaged by the formation of the lesion. CD is treated by strict adherence to a gluten-free diet, which usually results in complete remission [6]. While barley, rye, and wheat undoubtedly contain immunogenic proteins, there is some evidence that oats, prepared in wheat-free facilities, may be safe for consumption by celiac sufferers [7].

Refractory CD has been postulated in a minority of cases where the gluten-free diet is unsuccessful [8]. However, it is difficult to distinguish true refractory disease from noncompliance to the rigid requirements of the diet. Recent literature has also considered the possibility of non-celiac gluten or wheat intolerance [9, 10], which is beyond the scope of this chapter.

2   Historical Aspects

There is evidence for the use of grains such as wild wheat and barley dating to 23,000 years ago in the upper Paleolithic [11, 12]. However, the use of cereals did not become widespread until after the Neolithic revolution of approximately 10,000 years ago. While natural selection may have played a role at this stage, many of the alleles associated with CD risk may have been maintained at high frequency due to their ability to confer other beneficial traits, such as resistance to bacterial infection [13]. Archaeological evidence of probable CD has been identified in 2000-year-old human remains from Italy [14]. Interestingly, there is some evidence that einkorn (*Triticum monococcum*), the earliest cultivated wheat, may be less toxic to celiac sufferers than more modern varieties [15].

CD has been recognized since ancient times. It was first described in the first century BC by the Greek physician Aretaeus of Cappadocia, whose works were translated in the 1800s [16]. Aretaeus identified CD as an affliction of later life, most commonly affecting women. The physician Samuel Gee gave the first modern description of the condition in 1888, building upon Aretaeus' observations.

However, he primarily observed the condition in infants, and considered it a disease of childhood. The "classical" picture of CD—occurring in the young, presenting with characteristic abdominal symptoms, diarrhea, and "failure to thrive"—owes itself to Gee's observations at this time [17].

A dietary, specifically carbohydrate, component to CD was long suspected. The first treatments for CD pre-date full understanding of the etiology, for example the "banana diet" [18]. However, it was not until the 1940s that the physician Wilhelm Dicke identified the ingestion of wheat as the environmental trigger, aided by the observation that reduced morbidity from CD coincided with the shortage of wheat during the Dutch Hongerwinter of 1944 [19].

3   Evolving Understanding of Onset and Prevalence

The view of CD, as described by Gee, is one of a rare illness of childhood, affecting individuals of European descent. As recently as 1985, estimates of incidence of CD placed the population frequency at 0.1 %. Even at this time however, the classical view of the disease as an affliction of childhood was beginning to be questioned, with the Celiac Society noting an increase in age of diagnosis amongst its members [20].

There is some evidence that the age of onset of adult CD may follow a bimodal distribution, with an initial peak in the fourth decade of life (mostly women) and a second, smaller peak in the sixth or seventh decade of life (predominantly men) [21]. The apparent increase in incidence amongst the older population may be due to improved screening and diagnostic techniques, leading to the recognition of CD in cases where it would previously have gone unnoticed or misdiagnosed. Additionally, individuals with "silent," or asymptomatic, disease may convert to a more aggressive phenotype with age. Increased recognition of nonclassical symptoms (active, silent, latent, and potential CD) may aid in diagnosis [22].

The prevalence of asymptomatic CD is unknown, and indeed known cases may constitute the "tip of the iceberg" [23]. Definitive diagnosis requires endoscopic biopsy, which is invasive and not conducive to population screening. Serological tests which assay for the presence of anti-tTG autoantibodies have also been developed as a diagnostic method. Many researchers have screened populations using these serological assays, which can give good estimates of the true population prevalence. However, estimates of prevalence based on serology alone, in the absence of endoscopic confirmation and HLA genotyping, will contain a proportion of false positives.

Historically, CD risk has been most comprehensively described in populations of European descent [1]. However, the DQ2 and DQ8 variants are widespread in worldwide populations, and the condition was once considered rare in some populations where it is now known to be present. Therefore, there is a precedent for underdiagnosis of CD, and future studies may reveal a more widespread distribution, particularly as the consumption of wheat increases, as observed in China [24].

4   The Immunology of Celiac Disease

CD is driven by aberrant immune responses to dietary gluten in genetically predisposed individuals. Therefore, in addition to genetic and environmental components, the immune system plays an important role in CD pathogenesis.

The immune system is divided into innate and adaptive arms. The innate immune system is comprised of first-line defence barriers and rapidly responding immune cells activated by conserved molecular patterns. The adaptive immune system, comprised of T cells and B cells, can take days to mount a response but specifically recognizes peptide antigens and can develop immunological memory. Antigen-presenting cells bridge the innate and adaptive immune systems by processing and presenting peptide antigens via MHC molecules to stimulate T cell responses. In the context of CD, antigen-presenting cells present deamidated gliadin peptides bound to MHC molecules HLA-DQ2 or HLA-DQ8 to activate gluten-specific T cells.

One of the early hallmarks of CD is an influx of activated lymphocytes into the small intestine, concomitant with villous atrophy [25, 26]. Infiltrating T cells show memory and cytotoxic phenotypes and are thought to drive mucosal damage in response to gluten peptides [27, 28]. It has been postulated that gluten-specific CD4+ T cells found in the lamina propria activate intraepithelial CD8+ T cells and B cells via cross-priming and cytokine production, which further drives inflammation and tissue damage [29, 30]. Cytokines, such as interferon-γ (IFN-γ), interleukin-15 (IL-15), and IL-21 in particular, have been implicated in CD pathogenesis [31, 32]. IFN-γ, a pro-inflammatory T helper 1 (T_H1) type cytokine, increases intestinal barrier permeability and drives cytotoxic CD8+ T cell activation, whereas IL-15 further drives T_H1 functions and cytotoxic T cell survival.

Activated B cells produce antibodies specific for gluten peptides and tissue transglutaminase, the latter providing a useful diagnostic test for CD [33]. Other types of infiltrating immune cells have also been described in the celiac small intestine. Gamma delta ($\gamma\delta$) T cells, particularly the Vδ1 subtype, are significantly enriched in the human small intestinal epithelium, persisting even after elimination of gluten from the diet [34, 35]. The precise role of these cells in CD remains unclear, but immunoregulatory and tissue repair functions have been described, suggesting that Vδ1 cells may aid the restoration and maintenance of the small intestinal epithelium [36, 37].

Chronic activation of T cells can result in lymphomagenesis in a small number of CD patients, leading to the development of aggressive enteropathy-associated T cell lymphomas (EATLs) with poor prognosis [38, 39].

5   Molecular Basis

Sollid et al. [40] identified the HLA-DQ variants as the main disease susceptibility factor. The contribution of these variants to celiac disease risk is considerably greater than any of the other known genetic risk factors.

Gluten, found in the endosperm of wheat, barley, and rye, is composed of various subunits. One class of these subunits, the gliadins, is capable of triggering an immunogenic response in CD [6, 41]. Gliadin is composed of more than a hundred components, which can be classified into four main types, termed ω5-, ω1,2-, α/β-, and γ-gliadins. Multiple gliadin-derived epitopes are immunogenic and toxic to CD sufferers [42]. Storage proteins with similar amino acid composition and toxicity to celiac sufferers have been identified in barley and rye (hordeins and secalines, respectively) [43].

Although many gluten-derived epitopes are immunogenic, they display a hierarchy of immunogenicity. A peptide of 33 amino acids (residues 57–89) derived from an α-gliadin fraction appears to be immune-dominant, properties attributable to its proline and glutamine-rich peptide structure. The density of proline residues increases the resistance of the peptide to gastrointestinal proteolysis, both in CD and in unaffected individuals. In addition, the resulting left-handed helical conformation strengthens binding to HLA-DQ2 and HLA-DQ8 molecules. tTG-mediated deamidation of gliadin peptides results in further enhanced immunogenicity [44]. The criteria for epitope binding to DQ2 or DQ8 are well recognized, with both DQ2 and DQ8 dimers exhibiting preferences for negatively charged residues at positions within the core peptide-binding groove [45, 46]. These negatively charged residues are the result of the deamidation of glutamine residues by tTG [47]. Therefore, the endogenous enzyme tTG, implicated in CD due to the presence of autoantibodies, is the catalyst for the deamidation and subsequent increase in epitope affinity for HLA-DQ2 and HLA-DQ8 binding sites, enhancing gliadin-specific T cell activity [48].

Tellingly, tTG is highly expressed in the small intestine [49] and upregulated in inflammation. In CD, excess tTG favors the association of tTG/tTG-gliadin complexes with collagens, acting to localize the gliadin to the lamina propria, favoring the progression of the disease [50, 51].

In essence then, the modern picture of CD is of an autoimmune disorder with a strong, necessary environmental component (gliadin) which acts as a substrate for the endogenous enzyme tTG and results in the production of autoantibodies. However, we must also ask why some individuals are affected by the disease and others are not, a question which leads us towards discussion of the genetic underpinnings of the condition.

6   Genetic Susceptibility

Early studies of CD focussed on the HLA, located on the short arm of chromosome 6 (6p21). Population genetic variation at this hypervariable region of the genome can be assayed using serological and, more recently, PCR-based assays. The major celiac susceptibility factor HLA-DQ2, an ancient and widespread variant, is located here, in the MHC Class II region [40].

In addition, there has been much research which attempted to identify additional risk loci within the HLA. The Class III region, which houses many plausible candidate genes such as *TNFα*, has attracted much attention [52]. However, the HLA region exhibits long range linkage disequilibrium and high levels of polymorphism, probably maintained by balancing natural selection. While most regions of the human genome coalesce to a most recent common ancestor (MRCA) within the last 200,000 years, the mean time to MRCA for HLA alleles may be in excess of 20 million years [53, 54], an unusual time depth for the human genome. Indeed, Abi-Rachad et al. showed that many of the common haplotypes present in non-African populations may have originated in archaic humans, namely Neanderthals and Denisovans [55]. For these reasons, fine mapping of genetic susceptibility within the HLA region has proven difficult.

Linkage analysis has been applied to CD. While this study design can be powerful for the detection of large effects, it has met with limited success in studies of complex diseases, due to the tendency of these diseases to have multiple risk factors of small to moderate effect. Nonetheless, linkage analyses have given strong suggestions of a risk factor on chromosome 5q31 [56, 57], though no evidence for this has been seen in candidate gene [58, 59] or genome-wide analyses.

Candidate gene studies have also uncovered a number of effects. The co-stimulator region on chromosome 2q33 [60], originally identified in a candidate gene study, has been replicated many times, though paradoxically numerous other studies have failed to replicate the effect. Interestingly, worldwide haplotype structure in this region shows extreme geographical population structure [61], a fact which might explain some of the difficulty in replicating the association.

Genome-wide association studies (GWAS) have provided a hypothesis-free approach to elucidate the etiology of complex diseases. In addition to identifying genetic risk, this approach may also identify potential drug targets. The advent of GWAS greatly expanded the number of robust genetic associations for CD [62–64]. To date, 43 non-HLA-associated loci have been identified via this approach [65–67], which together with the HLA variants explain up to 54 % of the population variability associated with CD risk.

The mechanism by which many disease-associated variants exert a functional effect is not completely understood. However, correlation between noncoding variation and gene expression has been observed in CD [68], and different patterns of gene expression have been observed in CD patients compared to controls, both in peripheral blood monocytes and in biopsies [69]. In addition, a suggestion of nonenzymatic functions of trypsin and pepsin has been revealed by gene expression profiling of the effect of gliadin on epithelial cells [70]. Studies of this kind lead to a more complete molecular characterization of CD and may functionally explain how intergenic polymorphisms might influence phenotype as expression quantitative trait loci.

7   Specific Requirements for CD Research

As with any complex condition, the proper diagnosis and characterization of the sample population is crucial. In the case of CD, this strictly requires endoscopic biopsy, seropositivity for anti-tTG, and preferably confirmation of HLA-DQ2 or DQ8 carrier status. However, guaranteeing this level of proof is not always trivial for large studies. In addition, ethical permission and informed consent are required for all studies involving patient samples.

Patients can often be recruited as they attend clinic. The timing of sampling for functional studies should preferably be before the patient begins the gluten-free diet, bearing in mind that symptoms may begin to wane quickly, and anecdotal evidence suggests that some patients begin on the diet prior to definitive diagnosis. In addition, it may be necessary to target the right tissue for the question at hand. Many questions may be addressed by relatively noninvasive collection of peripheral blood. However, more in-depth analyses may require the sampling, proper handling, and storage of biopsies from gluten-challenged patients. A key consideration here is that diagnostic and management approaches may differ geographically and temporally, which may present a challenge for collaborative studies.

For large-scale genetic analyses, sample size and genome coverage are key. With the growth of high-throughput analyses, many large datasets are currently being made available to the research community, a development which affords considerable opportunity for bioinformatic analysis and fine mapping. As the body of existing data grows, so too do the opportunities to perform focused analysis on selected genomic regions, genes involved in common pathways, or genes whose products are typically co-expressed in cells relevant to disease.

8   Future Prospects

Next-generation sequencing technology has allowed many individual human genomes to be sequenced. This number is set to increase further as the associated costs fall, and indeed the genomes of many ancient individuals have been sequenced [71, 72]. At the same time, evidence of CD can be identified in ancient human remains [14]. These approaches may well shed some light on the transition of ancient populations to agriculture and wheat consumption. The bioinformatic tools required for the analysis of these datasets will become increasingly important as more exome and genome sequences become available.

A succession of recent genome-wide association studies has been highly successful in identifying CD risk variants. At present, up to 54 % of the variance in CD risk can be explained by known variants, including HLA. Further research will no doubt explain an even greater proportion of the disease risk, as well as refine the methods used to calculate it. It is clear that there is still much to learn.

The human genome contains numerous copy number variants, some of which have been associated with autoimmune disease, for example the deletion of late cornified envelope genes *LCE3C* and *LCE3B* in psoriasis [73]. Genome-wide analyses of neuropsychiatric conditions have focused much more closely on analysis of copy number variants and rare variants, which could potentially explain some of the missing heritability, but are by their nature difficult to study [74]. While there is some evidence that rare variants may play a role in CD susceptibility [66], no copy number variant risk factors have as yet been identified.

Given the long-term, chronic nature of CD, it appears likely that some epigenetic factors come into play in the development of the disease. In this regard, a recent study has demonstrated a role for DNA methylation in CD [75], and there has been some suggestion of parent of origin effects in genetic susceptibility [76]. Epigenetic effects have, therefore, received little attention in CD research and may afford a fruitful avenue for further investigation.

In conclusion, the short time elapsed since the introduction of genome-wide analysis has seen a revolution in our understanding of CD genetic risk. In the meantime, genomic and other high-throughput technologies have continued to develop at a rapid pace. The coming years will likely see further developments and insights, both in terms of understanding the disease and translating the discoveries into patient care.

Acknowledgements

Thanks to Derek Doherty and Emma Quinn for reading and commenting on this manuscript, and to Eleisa Heron for useful discussion. Anthony W. Ryan acknowledges funding from the Royal City of Dublin Hospital Trust.

References

1. Fasano A, Catassi C (2012) Clinical practice. Celiac disease. N Engl J Med 367:2419–26. doi:10.1056/NEJMcp1113994

2. Reunala TL (2001) Dermatitis herpetiformis. Clin Dermatol 19:728–36

3. Sollid LM (2002) Celiac disease: dissecting a complex inflammatory disorder. Nat Rev Immunol 2:647–55. doi:10.1038/nri885

4. Abraham G, Tye-Din JA, Bhalala OG et al (2014) Accurate and robust genomic prediction of celiac disease using statistical learning. PLoS Genet 10:e1004137. doi:10.1371/journal.pgen.1004137

5. Kneepkens CMF, von Blomberg BME (2012) Clinical practice : coeliac disease. Eur J Pediatr 171:1011–21. doi:10.1007/s00431-012-1714-8

6. Di Sabatino A, Corazza GR (2009) Coeliac disease. Lancet 373:1480–1493. doi:10.1016/S0140-6736(09)60254-3

7. Cooper SEJ, Kennedy NP, Mohamed BM et al (2013) Immunological indicators of coeliac disease activity are not altered by long-term oats challenge. Clin Exp Immunol 171:313–8. doi:10.1111/cei.12014

8. O'Shea U, Abuzakouk M, O'Morain C et al (2008) Investigation of molecular markers in the diagnosis of refractory coeliac disease in a large patient cohort. J Clin Pathol 61:1200–2. doi:10.1136/jcp.2008.058404

9. Biesiekierski JR, Newnham ED, Irving PM et al (2011) Gluten causes gastrointestinal symptoms in subjects without celiac disease: a double-blind randomized placebo-controlled trial. Am J Gastroenterol 106:508–14. doi:10.1038/ajg.2010.487, quiz 515

10. Carroccio A, Rini G, Mansueto P (2014) Non-celiac wheat sensitivity is a more appropriate label than non-celiac gluten sensitivity. Gastroenterology 146:320–1. doi:10.1053/j.gastro.2013.08.061

11. Piperno DR, Weiss E, Holst I, Nadel D (2004) Processing of wild cereal grains in the Upper Palaeolithic revealed by starch grain analysis. Nature 430:670–3. doi:10.1038/nature02734

12. Weiss E, Wetterstrom W, Nadel D, Bar-Yosef O (2004) The broad spectrum revisited: evidence from plant remains. Proc Natl Acad Sci U S A 101:9551–5. doi:10.1073/pnas.0402362101

13. Zhernakova A, Elbers CC, Ferwerda B et al (2010) Evolutionary and functional analysis of celiac risk loci reveals SH2B3 as a protective factor against bacterial infection. Am J Hum Genet 86:970–7. doi:10.1016/j.ajhg.2010.05.004

14. Scorrano G, Brilli M, Martínez-Labarga C et al (2014) Palaeodiet reconstruction in a woman with probable celiac disease: a stable isotope analysis of bone remains from the archaeological site of Cosa (Italy). Am J Phys Anthropol 154:349–56. doi:10.1002/ajpa.22517

15. Pizzuti D, Buda A, D'Odorico A et al (2006) Lack of intestinal mucosal toxicity of Triticum monococcum in celiac disease patients. Scand J Gastroenterol 41:1305–11. doi:10.1080/00365520600699983

16. Dowd B, Walker-Smith J (1974) Samuel Gee, Aretaeus, and the coeliac affection. Br Med J 2:45–7

17. Gee SJ (1888) On the coeliac affection. St Bartholomews Hosp Rep 24:17–20

18. Haas SV (1924) The value of the banana in the treatment of celiac disease. Am J Dis Child 28:421–37. doi:10.1001/archpedi.1924.04120220017004

19. Van Berge-Henegouwen GP, Mulder CJ (1993) Pioneer in the gluten free diet: Willem-Karel Dicke 1905-1962, over 50 years of gluten free diet. Gut 34:1473–5

20. Langman MJ, McConnell TH, Spiegelhalter DJ, McConnell RB (1985) Changing patterns of coeliac disease frequency: an analysis of Coeliac Society membership records. Gut 26:175–8

21. Swinson CM, Levi AJ (1980) Is coeliac disease underdiagnosed? Br Med J 281:1258–60

22. Ferguson A, Arranz E, O'Mahony S (1993) Clinical and pathological spectrum of coeliac disease–active, silent, latent, potential. Gut 34:150–1

23. Catassi C, Fabiani E, Rätsch IM et al (1996) The coeliac iceberg in Italy. A multicentre antigliadin antibodies screening for coeliac disease in school-age subjects. Acta Paediatrica, 85:29–35. doi:10.1111/j.1651-2227.1996.tb14244.x

24. Yuan J, Gao J, Li X et al (2013) The tip of the "celiac iceberg" in China: a systematic review and meta-analysis. PLoS One 8:e81151. doi:10.1371/journal.pone.0081151

25. Ferguson A, Murray D (1971) Quantitation of intraepithelial lymphocytes in human jejunum. Gut 12:988–94

26. Halstensen TS, Brandtzaeg P (1993) Activated T lymphocytes in the celiac lesion: non-proliferative activation (CD25) of CD4+ alpha/beta cells in the lamina propria but proliferation (Ki-67) of alpha/beta and gamma/delta cells in the epithelium. Eur J Immunol 23:505–10. doi:10.1002/eji.1830230231

27. Buri C, Burri P, Bähler P et al (2005) Cytotoxic T cells are preferentially activated in the duodenal epithelium from patients with florid coeliac disease. J Pathol 206:178–85. doi:10.1002/path.1773

28. Lundin KE, Scott H, Hansen T et al (1993) Gliadin-specific, HLA-DQ(alpha 1*0501, beta 1*0201) restricted T cells isolated from the small intestinal mucosa of celiac disease patients. J Exp Med 178:187–96

29. Meresse B, Malamut G, Cerf-Bensussan N (2012) Celiac disease: an immunological jigsaw. Immunity 36:907–19. doi:10.1016/j.immuni.2012.06.006

30. Sollid LM, Jabri B (2013) Triggers and drivers of autoimmunity: lessons from coeliac disease. Nat Rev Immunol 13:294–302. doi:10.1038/nri3407

31. Fina D, Sarra M, Caruso R et al (2008) Interleukin 21 contributes to the mucosal T helper cell type 1 response in coeliac disease. Gut 57:887–92. doi:10.1136/gut.2007.129882

32. Di Sabatino A, Ciccocioppo R, Cupelli F et al (2006) Epithelium derived interleukin 15 regulates intraepithelial lymphocyte Th1 cytokine production, cytotoxicity, and survival in coeliac disease. Gut 55:469–77. doi:10.1136/gut.2005.068684

33. Dieterich W, Ehnis T, Bauer M et al (1997) Identification of tissue transglutaminase as the autoantigen of celiac disease. Nat Med 3:797–801

34. Halstensen TS, Scott H, Brandtzaeg P (1989) Intraepithelial T cells of the TcR gamma/delta + CD8- and V delta 1/J delta 1+ phenotypes are increased in coeliac disease. Scand J Immunol 30:665–72

35. Dunne MR, Elliott L, Hussey S et al (2013) Persistent changes in circulating and intestinal γδ T cell subsets, invariant natural killer T cells and mucosal-associated invariant T cells in children and adults with coeliac disease. PLoS One 8:e76008. doi:10.1371/journal.pone.0076008

36. Bhagat G, Naiyer AJ, Shah JG et al (2008) Small intestinal CD8 + TCRgammadelta + NKG2A+ intraepithelial lymphocytes have attributes of regulatory cells in patients with celiac disease. J Clin Invest 118:281–93. doi:10.1172/JCI30989

37. Toulon A, Breton L, Taylor KR et al (2009) A role for human skin-resident T cells in wound healing. J Exp Med 206:743–50. doi:10.1084/jem.20081787

38. O'Farrelly C, Feighery C, O'Briain DS et al (1986) Humoral response to wheat protein in patients with coeliac disease and enteropathy associated T cell lymphoma. Br Med J (Clin Res Ed) 293:908–10

39. Cellier C, Delabesse E, Helmer C et al (2000) Refractory sprue, coeliac disease, and enteropathy-associated T-cell lymphoma. French Coeliac Disease Study Group. Lancet 356:203–8

40. Sollid LM, Markussen G, Ek J et al (1989) Evidence for a primary association of celiac disease to a particular HLA-DQ alpha/beta heterodimer. J Exp Med 169:345–50

41. Wieser H (1996) Relation between gliadin structure and coeliac toxicity. Acta Paediatr Suppl 412:3–9

42. Kagnoff MF (2007) Celiac disease: pathogenesis of a model immunogenetic disease. J Clin Invest 117:41–9. doi:10.1172/JCI30253

43. Wieser H (1995) The precipitating factor in coeliac disease. Baillieres Clin Gastroenterol 9:191–207

44. Shan L, Molberg Ø, Parrot I et al (2002) Structural basis for gluten intolerance in celiac sprue. Science 297:2275–9. doi:10.1126/science.1074129

45. Tollefsen S, Arentz-Hansen H, Fleckenstein B et al (2006) HLA-DQ2 and -DQ8 signatures of gluten T cell epitopes in celiac disease. J Clin Invest 116:2226–36. doi:10.1172/JCI27620

46. Vartdal F, Johansen BH, Friede T et al (1996) The peptide binding motif of the disease associated HLA-DQ (alpha 1* 0501, beta 1* 0201) molecule. Eur J Immunol 26:2764–72. doi:10.1002/eji.1830261132

47. Qiao S-W, Bergseng E, Molberg O et al (2005) Refining the rules of gliadin T cell epitope binding to the disease-associated DQ2 molecule in celiac disease: importance of proline spacing and glutamine deamidation. J Immunol 175:254–61

48. Van de Wal Y, Kooy Y, van Veelen P et al (1998) Selective deamidation by tissue transglutaminase strongly enhances gliadin-specific T cell reactivity. J Immunol 161:1585–8

49. Fleckenstein B, Molberg Ø, Qiao S-W et al (2002) Gliadin T cell epitope selection by tissue transglutaminase in celiac disease. Role of enzyme specificity and pH influence on the transamidation versus deamidation process. J Biol Chem 277:34109–16. doi:10.1074/jbc.M204521200

50. Dieterich W, Esslinger B, Trapp D et al (2006) Cross linking to tissue transglutaminase and collagen favours gliadin toxicity in coeliac disease. Gut 55:478–84. doi:10.1136/gut.2005.069385

51. Griffin M, Casadio R, Bergamini CM (2002) Transglutaminases: nature's biological glues. Biochem J 368:377–96. doi:10.1042/BJ20021234

52. McManus R, Moloney M, Borton M et al (1996) Association of celiac disease with microsatellite polymorphisms close to the tumor necrosis factor genes. Hum Immunol 45:24–31

53. Otting N, de Groot NG, Doxiadis GGM, Bontrop RE (2002) Extensive Mhc-DQB variation in humans and non-human primate species. Immunogenetics 54:230–9. doi:10.1007/s00251-002-0461-9

54. Otting N, Kenter M, van Weeren P et al (1992) Mhc-DQB repertoire variation in hominoid and Old World primate species. J Immunol 149:461–70

55. Abi-Rached L, Jobin MJ, Kulkarni S et al (2011) The shaping of modern human immune systems by multiregional admixture with archaic humans. Science 334:89–94. doi:10.1126/science.1209202

56. Greco L, Babron MC, Corazza GR et al (2001) Existence of a genetic risk factor on chromosome 5q in Italian coeliac disease families. Ann Hum Genet 65:35–41

57. Adamovic S, Amundsen SS, Lie BA et al (2008) Fine mapping study in Scandinavian families suggests association between coeliac disease and haplotypes in chromosome region 5q32. Tissue Antigens 71:27–34. doi:10.1111/j.1399-0039.2007.00955.x

58. Ryan AW, Thornton JM, Brophy K et al (2004) Haplotype variation at the IBD5/SLC22A4 locus (5q31) in coeliac disease in the Irish population. Tissue Antigens 64:195–8. doi:10.1111/j.1399-0039.2004.00251.x

59. Ryan AW, Thornton JM, Brophy K et al (2005) Chromosome 5q candidate genes in coeliac disease: genetic variation at IL4, IL5, IL9, IL13, IL17B and NR3C1. Tissue Antigens 65:150–5. doi:10.1111/j.1399-0039.2005.00354.x

60. Holopainen P, Arvas M, Sistonen P et al (1999) CD28/CTLA4 gene region on chromosome 2q33 confers genetic susceptibility to celiac disease. A linkage and family-based association study. Tissue Antigens 53:470–5

61. Butty V, Roy M, Sabeti P et al (2007) Signatures of strong population differentiation shape extended haplotypes across the human CD28, CTLA4, and ICOS costimulatory genes. Proc Natl Acad Sci U S A 104:570–5. doi:10.1073/pnas.0610124104

62. Van Heel DA, Franke L, Hunt KA et al (2007) A genome-wide association study for celiac disease identifies risk variants in the region harboring IL2 and IL21. Nat Genet 39:827–9. doi:10.1038/ng2058

63. Hunt KA, Zhernakova A, Turner G et al (2008) Newly identified genetic risk variants for celiac disease related to the immune response. Nat Genet 40:395–402. doi:10.1038/ng.102

64. Dubois PCA, Trynka G, Franke L et al (2010) Multiple common variants for celiac disease influencing immune gene expression. Nat Genet 42:295–302. doi:10.1038/ng.543

65. Garner C, Ahn R, Ding YC et al (2014) Genome-wide association study of celiac disease in North America confirms FRMD4B as new celiac locus. PLoS One 9:e101428. doi:10.1371/journal.pone.0101428

66. Trynka G, Hunt KA, Bockett NA et al (2011) Dense genotyping identifies and localizes multiple common and rare variant association signals in celiac disease. Nat Genet 43:1193–201. doi:10.1038/ng.998

67. Coleman C, Quinn EM, Ryan AW, et al. (2015) Common polygenic variation in coeliac

disease and confirmation of ZNF335 and NIFA as disease susceptibility loci. Eur J Hum Genet (in press)

68. Heap GA, Trynka G, Jansen RC et al (2009) Complex nature of SNP genotype effects on gene expression in primary human leucocytes. BMC Med Genomics 2:1. doi:10.1186/1755-8794-2-1

69. Galatola M, Izzo V, Cielo D et al (2013) Gene expression profile of peripheral blood monocytes: A step towards the molecular diagnosis of celiac disease? PLoS One 8:e74747. doi:10.1371/journal.pone.0074747

70. Parmar A, Greco D, Venäläinen J et al (2013) Gene expression profiling of gliadin effects on intestinal epithelial cells suggests novel non-enzymatic functions of pepsin and trypsin. PLoS One 8:e66307. doi:10.1371/journal.pone.0066307

71. Gamba C, Jones ER, Teasdale MD et al (2014) Genome flux and stasis in a five millennium transect of European prehistory. Nat Commun 5:5257. doi:10.1038/ncomms6257

72. Lazaridis I, Patterson N, Mittnik A et al (2014) Ancient human genomes suggest three ancestral populations for present-day Europeans. Nature 513:409–13. doi:10.1038/nature13673

73. Riveira-Munoz E, He S-M, Escaramís G et al (2011) Meta-analysis confirms the LCE3C_LCE3B deletion as a risk factor for psoriasis in several ethnic groups and finds interaction with HLA-Cw6. J Invest Dermatol 131:1105–9. doi:10.1038/jid.2010.350

74. Rees E, Moskvina V, Owen MJ et al (2011) De novo rates and selection of schizophrenia-associated copy number variants. Biol Psychiatry 70:1109–14. doi:10.1016/j.biopsych.2011.07.011

75. Fernandez-Jimenez N, Castellanos-Rubio A, Plaza-Izurieta L et al (2014) Coregulation and modulation of NFκB-related genes in celiac disease: uncovered aspects of gut mucosal inflammation. Hum Mol Genet 23:1298–310. doi:10.1093/hmg/ddt520

76. Megiorni F, Mora B, Bonamico M et al (2008) HLA-DQ and susceptibility to celiac disease: evidence for gender differences and parent-of-origin effects. Am J Gastroenterol 103:997–1003. doi:10.1111/j.1572-0241.2007.01716.x

Chapter 2

Celiac Disease: Diagnosis

Greg Byrne and Conleth F. Feighery

Abstract

Historically the diagnosis of celiac disease has relied upon clinical, serological, and histological evidence. In recent years the use of sensitive serological methods has meant an increase in the diagnosis of celiac disease. The heterogeneous nature of the disorder presents a challenge in the study and diagnosis of the disease with patients varying from subclinical or latent disease to patients with overt symptoms. Furthermore the related gluten-sensitive disease dermatitis herpetiformis, while distinct in some respects, shares clinical and serological features with celiac disease. Here we summarize current best practice for the diagnosis of celiac disease and briefly discuss newer approaches. The advent of next-generation assays for diagnosis and newer clinical protocols may result in more sensitive screening and ultimately the possible replacement of the intestinal biopsy as the gold standard for celiac disease diagnosis.

Key words Celiac, Diagnosis, Symptoms, Autoantibodies

1 When to Suspect Celiac Disease: Signs and Symptoms

Celiac disease was originally considered an exclusively pediatric disorder, but it is now appreciated that this condition frequently presents in adulthood, even in the seventh and eight decade of life [1]. It is likely that the disease actually developed late in life in older patients rather than being always present in a latent form. In younger patients, especially in childhood, symptoms may be more obvious and are often gastrointestinal in nature. Hence pediatric patients may have a history of chronic diarrhea, abdominal distension, loss of appetite, and failure to thrive [2]. Older patients may have much less specific symptoms such as lack of energy, and this may be caused by anemia. Malabsorption is the key pathological consequence of celiac disease and anemia results from iron, folic acid, or even vitamin B12 deficiency [3]. Patients may also develop osteoporosis caused by vitamin D and calcium malabsorption [4]. Because of the protean manifestations of celiac disease, it is sensible to keep this diagnosis in mind in many clinical situations. For example, individuals with mild anemia,

Anthony W. Ryan (ed.), *Celiac Disease: Methods and Protocols*, Methods in Molecular Biology, vol. 1326,
DOI 10.1007/978-1-4939-2839-2_2, © Springer Science+Business Media New York 2015

with autoimmune endocrine diseases (such as thyroid disease or diabetes), with oral ulcers or unexplained weight loss can reasonably be investigated for celiac disease. The uncommon skin disease, dermatitis herpetiformis, may also be a presenting complaint with a typical intensely itchy vesicular rash on extensor surfaces [5].

2 HLA Typing

The vast majority (95 %) of patients with celiac disease express the HLA-DQ2 heterodimer in *cis* or *trans* and HLA-DQ8 is present in most of the remaining patients [6]. HLA-DQ2 and HLA-DQ8 have been estimated to contribute ~35 % of the risk of developing celiac disease [7]. The strongest association is with HLA-DQA*0501, DQB*0201 (termed HLADQ2.5) which, when inherited in a homozygous manner, is associated with a fivefold increased risk for the development of celiac disease when compared to individuals heterozygous for HLA-DQ2.5. Another HLA-DQ2 variant exists termed HLA-DQ2.2 (DQA*0201, DQB*0202), which has an almost identical peptide-binding motif to HLA-DQ2.5 but interestingly does not predispose to celiac disease [8]. Given the prevalence of these HLA types in Western populations (~40 %), HLA typing has very poor positive predictive value when considered for celiac disease diagnosis [9]. However, the absence of both HLA-DQ2 and HLA-DQ8 makes diagnosis of celiac disease highly unlikely. HLA typing of patients has been included as a useful tool to exclude celiac disease in the ESPGHAN guidelines for diagnosis of celiac disease [2].

3 Serology

3.1 Anti-Gliadin Antibodies

Anti-gliadin antibody detection was the serological method of choice for investigation of celiac disease but exhibited poor sensitivity and specificity [10]. As a consequence the use of IgG or IgA anti-gliadin antibodies is no longer recommended for use in the diagnosis of celiac disease [2, 11].

3.2 Endomysial Antibody (EMA) Test

The fluorescent detection of autoantibodies directed against endomysial antigens was developed in 1984 [12]. The endomysium is a layer of connective tissue that ensheathes muscle fibers [13]. The EMA test detects IgA class antibodies that bind to this connective tissue in monkey esophagus [14]. A positive result in the EMA test results in a characteristic fishnet pattern (Fig. 1). The specificity of the EMA test is as high as 98–100 % in experienced diagnostic laboratories [15]. The EMA test is a semi-quantitative method with serial dilution of patient serum being used to report a titer.

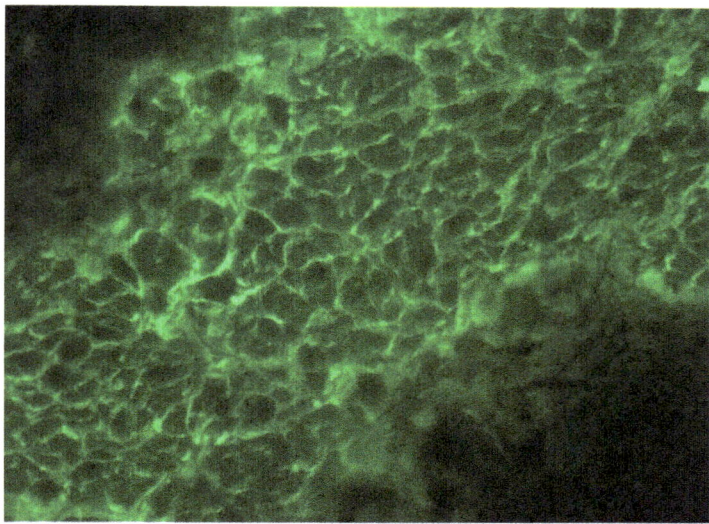

Fig. 1 Immunofluorescence pattern of endomysial antibodies (EMA). Celiac disease autoantibodies produce a distinctive fishnet pattern on monkey esophagus which is a very sensitive diagnostic marker

3.3 Tissue Transglutaminase Autoantibodies

The discovery by Dieterich et al. in 1997 that tissue transglutaminase (tTG) is the target autoantigen of endomysial autoantibodies [16] has facilitated the development of enzyme-linked methods to detect these autoantibodies. Subsequent demonstration that endomysium contains high levels of tissue transglutaminase explained the mechanism by which the EMA test is effective in celiac disease diagnosis [17]. tTG is the second member of a family of enzymes responsible for introducing inter-protein cross-links and is therefore often referred to as transglutaminase 2 (TG2) [18]. tTG plays an important role in celiac disease by catalyzing the site-specific deamidation of glutamine residues in gluten peptides, thereby increasing their affinity for the disease-associated HLA molecules and thus increasing T cell activation [19]. While detection of autoantibodies directed against tTG is the foundation of serological diagnosis of celiac disease, the origin of these autoantibodies and whether they play a direct role in the disease process remain debatable [20, 21].

IgA class antibodies against tissue transglutaminase are routinely detected by enzyme-linked immunosorbent assay. Patient levels are compared to a cutoff value determined by analyzing a healthy population and typically newly diagnosed, untreated patients have high levels of anti-tTG antibodies. The combination of anti-tTG and EMA tests results in positive and negative predictive values approaching 100 % [22]. The gluten-dependent nature of tissue transglutaminase antibodies has warranted their use as a marker of gluten-free diet compliance [23].

3.4 New Serological Assays

The detection of anti-gliadin antibodies has been replaced recently by deamidated gliadin peptide (DGP) assays. The antigens used in the assay replicate those generated in vivo by the deamidation of gliadin peptides by tissue transglutaminase. While it is unlikely that these assays will improve upon the EMA/tTG tests, IgG class DGPs have been shown to be useful in the diagnosis of celiac disease in patients <2 years of age [24].

Another new assay for diagnosis involves the detection of antibodies directed against DGP-tTG complexes. It is thought that these neoepitopes occur in vivo and are necessary for the development of the anti-tTG response to occur [21]. Current understanding of celiac disease pathogenesis suggests that these antibodies may arise prior to antibodies targeting tTG and DGP. Initial studies have shown diagnostic sensitivity (95 %) and specificity (97 %) comparable or greater than the current tTG and EMA paradigm [25].

3.5 IgA Deficiency and Celiac Disease

Celiac disease is associated with selective IgA deficiency with a prevalence of 3 % being reported [26, 27] compared to 0.17 % in healthy Caucasians. This unusually high prevalence must be taken into account in the case of negative tTG and EMA results where there is clinical suspicion of celiac disease. In these cases IgG class antibodies can be detected against the same autoantigens [27].

3.6 Use of Serological Assays in Other Gluten-Sensitive Diseases

Dermatitis herpetiformis is a related gluten-sensitive condition that manifests itself predominantly in the skin. The condition is characterized by a blistering, intensely pruritic papulovesicular rash typically located on the elbows, forearms, buttocks, knees, and scalp [5]. Histological examination of the cutaneous lesion reveals the presence of IgA deposits [28]. Gastrointestinal manifestations of gluten sensitivity tend to be milder or subclinical when compared to celiac disease [29]. While patients with dermatitis herpetiformis frequently have positive EMA and tTG results, the gold standard for diagnosis is the detection of granular IgA deposits in the dermal papillae. It has been recently discovered that these patients produce antibodies against epidermal transglutaminase (TG3) and the detection of these antibodies is a more specific serological marker of dermatitis herpetiformis [30].

4   Small Intestinal Biopsy

The histological examination of small bowel damage is central to celiac disease diagnosis [14, 31–33]. Because the lesion can be patchy, it is recommended that at least one biopsy be taken from the duodenal bulb and at least four biopsies from the second and third part of the duodenum [2]. While disease severity varies, the classical lesion is associated with villous atrophy and crypt hyperplasia. Other features include increased intraepithelial lymphocyte (IEL) numbers, elongated crypts, decreased villous/crypt ratio, increased enterocyte mitosis in the crypts, as well as infiltration of plasma cells, lymphocytes, mast cells, eosinophils, and basophils into the lamina propria. In addition, the epithelial cell brush border may be absent and these epithelial cells may appear abnormal [2]. The Marsh system is used to describe the extent of mucosal damage and ranges from Marsh 1 (increased IELs), Marsh 2 (increased IEL + crypt hyperplasia), Marsh 3a (partial villous atrophy), Marsh 3b (subtotal villous atrophy), and Marsh 3c (total villous atrophy). While a valuable tool in investigating celiac disease, the diagnosis should never be made on the basis of histology alone [34]. The improvement of histology after the adoption of a gluten-free diet is a valuable observation that confirms diagnosis.

The advancements in serological diagnosis have called into question the absolute necessity for intestinal biopsy to be performed for diagnosis of celiac disease. New guidelines published by ESPGHAN suggest that in cases where anti-tTG levels exceed ten times the cutoff, biopsy is unnecessary for diagnosis of pediatric celiac disease [2]. While some studies on adult cohorts have suggested that high titer anti-tTG results have a high positive predictive for villous atrophy [35], most clinicians still advocate the small intestinal biopsy as the gold standard [36]. A flow sheet detailing a typical workflow for celiac disease diagnosis is shown in Fig. 2.

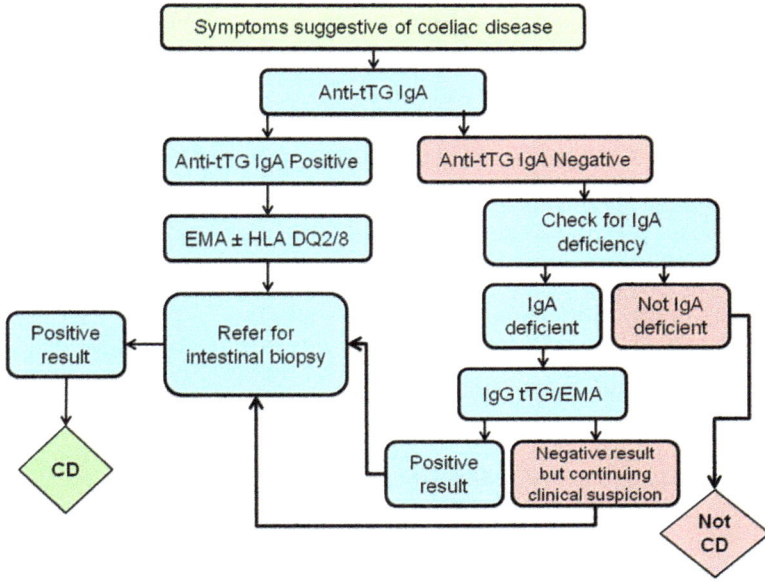

Fig. 2 Workflow for the diagnosis of celiac disease. Patients presenting with symptoms suggestive of celiac disease should be tested for IgA class anti-tTG antibodies. A positive result should be followed by an EMA test and intestinal biopsy. Care must be taken in the case of IgA deficiency where IgG tTG and EMA tests can be employed

5 Serological Markers of Intestinal Inflammation

The invasive, costly, and subjective nature of the histological examination of small intestinal architecture has motivated researchers to investigate the possibility of serological biomarkers of intestinal damage in celiac disease. In recent years the detection of anti-actin antibodies has been shown to correlate with the extent of intestinal inflammation in celiac disease. A multicenter study of anti-actin antibodies as a predictor of villous atrophy determined sensitivity and specificity of 80 and 85 % respectively [37]. Another candidate biomarker for intestinal inflammation in celiac disease is the soluble version of CD163. This marker is exclusively expressed on cells of the monocyte/macrophage lineage and acts as a scavenger receptor for hemoglobin/haptoglobin complexes. A soluble form of CD163 is shed from the cell upon activation [38], and levels have been shown to correlate with the Marsh stage of intestinal inflammation in celiac disease. Interestingly, elevated levels of this marker were not detected in the serum of patients with other gastrointestinal complaints, including inflammatory bowel disease, suggesting some degree of disease specificity [39].

References

1. Rashtak S, Murray JA (2009) Celiac disease in the elderly. Gastroenterol Clin North Am 38(3):433–446
2. Husby S, Koletzko S, Korponay-Szabo IR, Mearin ML, Phillips A, Shamir R, Troncone R, Giersiepen K, Branski D, Catassi C, Lelgeman M, Maki M, Ribes-Koninckx C, Ventura A, Zimmer KP (2012) European Society for Pediatric Gastroenterology, Hepatology, and Nutrition guidelines for the diagnosis of coeliac disease. J Pediatr Gastroenterol Nutr 54(1): 136–160
3. Kemppainen T, Uusitupa M, Janatuinen E, Jarvinen R, Julkunen R, Pikkarainen P (1995) Intakes of nutrients and nutritional status in coeliac patients. Scand J Gastroenterol 30(6):575–579
4. Hernandez L, Green PH (2006) Extraintestinal manifestations of celiac disease. Curr Gastroenterol Rep 8(5):383–389
5. Bolotin D, Petronic-Rosic V (2011) Dermatitis herpetiformis. Part I. Epidemiology, pathogenesis, and clinical presentation. J Am Acad Dermatol 64(6):1017–1024, quiz 1025-1016
6. Sollid LM, Thorsby E (1993) HLA susceptibility genes in celiac disease: genetic mapping and role in pathogenesis. Gastroenterology 105(3): 910–922
7. Hunt KA, Zhernakova A, Turner G et al (2008) Newly identified genetic risk variants for celiac disease related to the immune response. Nat Genet 40(4):395–402
8. Tjon JM, van Bergen J, Koning F (2010) Celiac disease: how complicated can it get? Immunogenetics 62(10):641–651
9. Kneepkens CM, von Blomberg BM (2012) Clinical practice: coeliac disease. Eur J Pediatr 171(7):1011–1021
10. Ascher H, Hahn-Zoric M, Hanson LA, Kilander AF, Nilsson LA, Tlaskalova H (1996) Value of serologic markers for clinical diagnosis and population studies of coeliac disease. Scand J Gastroenterol 31(1):61–67
11. Centre for Clinical Practice at NICE (UK) (2009) Coeliac disease: recognition and assessment of coeliac disease. National Institute for Health and Clinical Excellence (UK), London, UK
12. Chorzelski TP, Beutner EH, Sulej J, Tchorzewska H, Jablonska S, Kumar V, Kapuscinska A (1984) IgA anti-endomysium antibody. A new immunological marker of der-

matitis herpetiformis and coeliac disease. Br J Dermatol 111(4):395–402

13. Salmi TT, Collin P, Korponay-Szabo IR, Laurila K, Partanen J, Huhtala H, Kiraly R, Lorand L, Reunala T, Maki M, Kaukinen K (2006) Endomysial antibody-negative coeliac disease: clinical characteristics and intestinal autoantibody deposits. Gut 55:1746–1753

14. Feighery C (1999) Fortnightly review: coeliac disease. Br Med J 319(7204):236–239

15. Lewis NR, Scott BB (2006) Systematic review: the use of serology to exclude or diagnose coeliac disease (a comparison of the endomysial and tissue transglutaminase antibody tests). Aliment Pharmacol Ther 24(1):47–54

16. Dieterich W, Ehnis T, Bauer M, Donner P, Volta U, Riecken EO, Schuppan D (1997) Identification of tissue transglutaminase as the autoantigen of celiac disease. Nat Med 3(7):797–801

17. Brusco G, Muzi P, Ciccocioppo R, Biagi F, Cifone MG, Corazza GR (1999) Transglutaminase and coeliac disease: endomysial reactivity and small bowel expression. Clin Exp Immunol 118(3):371–375

18. Fesus L, Piacentini M (2002) Transglutaminase 2: an enigmatic enzyme with diverse functions. Trends Biochem Sci 27(10):534–539

19. Koning F, Gilissen L, Wijmenga C (2005) Gluten: a two-edged sword. Immunopathogenesis of celiac disease. Springer Semin Immunopathol 27(2):217–232

20. Caputo I, Barone MV, Martucciello S, Lepretti M, Esposito C (2009) Tissue transglutaminase in celiac disease: role of autoantibodies. Amino Acids 36(4):693–699

21. Sollid LM, Molberg O, McAdam S, Lundin KE (1997) Autoantibodies in coeliac disease: tissue transglutaminase–guilt by association? Gut 41(6):851–852

22. Schuppan D, Hahn EG (2001) IgA anti-tissue transglutaminase: setting the stage for coeliac disease screening. Eur J Gastroenterol Hepatol 13(6):635–637

23. Nachman F, Sugai E, Vazquez H, Gonzalez A, Andrenacci P, Niveloni S, Mazure R, Smecuol E, Moreno ML, Hwang HJ, Sanchez MI, Maurino E, Bai JC (2011) Serological tests for celiac disease as indicators of long-term compliance with the gluten-free diet. Eur J Gastroenterol Hepatol 23(6):473–480

24. Amarri S, Alvisi P, De Giorgio R, Gelli MC, Cicola R, Tovoli F, Sassatelli R, Caio G, Volta U (2013) Antibodies to deamidated gliadin peptides: an accurate predictor of coeliac disease in infancy. J Clin Immunol 33(5):1027–1030

25. Bizzaro N, Tozzoli R, Villalta D, Fabris M, Tonutti E (2012) Cutting-edge issues in celiac disease and in gluten intolerance. Clin Rev Allergy Immunol 42(3):279–287

26. Bottaro G, Cataldo F, Rotolo N, Spina M, Corazza GR (1999) The clinical pattern of subclinical/silent celiac disease: an analysis on 1026 consecutive cases. Am J Gastroenterol 94(3):691–696

27. Cataldo F, Lio D, Marino V, Picarelli A, Ventura A, Corazza GR (2000) IgG(1) antiendomysium and IgG antitissue transglutaminase (anti-tTG) antibodies in coeliac patients with selective IgA deficiency. Working Groups on Celiac Disease of SIGEP and Club del Tenue. Gut 47(3):366–369

28. van der Meer JB (1969) Granular deposits of immunoglobulins in the skin of patients with dermatitis herpetiformis. An immunofluorescent study. Br J Dermatol 81(7):493–503

29. Nakajima K (2012) Recent advances in dermatitis herpetiformis. Clin Dev Immunol 2012:914162

30. Borroni G, Biagi F, Ciocca O, Vassallo C, Carugno A, Cananzi R, Campanella J, Bianchi PI, Brazzelli V, Corazza GR (2013) IgA anti-epidermal transglutaminase autoantibodies: a sensible and sensitive marker for diagnosis of dermatitis herpetiformis in adult patients. J Eur Acad Dermatol Venereol 27(7):836–841

31. Green PH, Rostami K, Marsh MN (2005) Diagnosis of coeliac disease. Best Pract Res Clin Gastroenterol 19(3):389–400

32. Rodrigues AF, Jenkins HR (2008) Investigation and management of coeliac disease. Arch Dis Child 93(3):251–254

33. Walker-Smith JA, Guandalini S, Schmitz J, Shmerling D, Visakorpi J (1990) Revised criteria for diagnosis of coeliac disease. Report of Working Group of European Society of Paediatric Gastroenterology and Nutrition. Arch Dis Child 65(8):909–911

34. Collin P, Kaukinen K, Vogelsang H et al (2005) Antiendomysial and antihuman recombinant tissue transglutaminase antibodies in the diagnosis of coeliac disease: a biopsy-proven European multicentre study. Eur J Gastroenterol Hepatol 17(1):85–91

35. Vivas S, Ruiz de Morales JG, Riestra S et al (2009) Duodenal biopsy may be avoided when high transglutaminase antibody titers are present. World J Gastroenterol 15(38):4775–4780

36. Fernandez-Banares F, Alsina M, Modolell I et al (2012) Are positive serum-IgA-tissue-transglutaminase antibodies enough to diagnose coeliac disease without a small bowel biopsy? Post-test probability of coeliac disease. J Crohns Colitis 6(8):861–866

37. Carroccio A, Brusca I, Iacono G et al (2007) IgA anti-actin antibodies ELISA in coeliac disease: a multicentre study. Dig Liver Dis 39(9): 818–823

38. Droste A, Sorg C, Hogger P (1999) Shedding of CD163, a novel regulatory mechanism for a member of the scavenger receptor cysteine-rich family. Biochem Biophys Res Commun 256(1): 110–113

39. Daly A, Walsh C, Feighery C, O'Shea U, Jackson J, Whelan A (2006) Serum levels of soluble CD163 correlate with the inflammatory process in coeliac disease. Aliment Pharmacol Ther 24(3):553–559

Chapter 3

Generating Transgenic Mouse Models for Studying Celiac Disease

Josephine M. Ju, Eric V. Marietta, and Joseph A. Murray

Abstract

This chapter provides a brief overview of current animal models for studying celiac disease, with a focus on generating HLA transgenic mouse models. Human Leukocyte Antigen class II molecules have been a particular target for transgenic mice due to their tight association with celiac disease, and a number of murine models have been developed which had the endogenous MHC class II genes replaced with insertions of disease susceptible HLA class II alleles DQ2 or DQ8. Additionally, transgenic mice that overexpress interleukin-15 (IL-15), a key player in the inflammatory cascade that leads to celiac disease, have also been generated to model a state of chronic inflammation. To explore the contribution of specific bacteria in gluten-sensitive enteropathy, the nude mouse and rat models have been studied in germ-free facilities. These reductionist mouse models allow us to address single factors thought to have crucial roles in celiac disease. No single model has incorporated all of the multiple factors that make up celiac disease. Rather, these mouse models can allow the functional interrogation of specific components of the many stages of, and contributions to, the pathogenic mechanisms that will lead to gluten-dependent enteropathy. Overall, the tools for animal studies in celiac disease are many and varied, and provide ample space for further creativity as well as to characterize the complete and complex pathogenesis of celiac disease.

Key words Celiac, Gluten, Gliadin, In vivo, Animal model, Mouse, MHC II

1 Introduction

A variety of in vitro and in vivo models have been developed to study the innate and adaptive course of celiac disease, as well as its closely related skin manifestation, dermatitis herpetiformis. While epithelial cell lines and intestinal biopsy cultures allow us to look at celiac disease at a cellular level, animal models allow us to study how celiac disease works systemically and the way different organs are affected by the inflammatory immune response to gluten within the small intestine. To date, research has centered around mouse models, although some, mostly descriptive, work has been done in larger animals including dogs, monkeys, rabbits, rats, and even a horse with celiac associated antibodies and shortened villi which recovered following a 6-month gluten-free diet [1].

Anthony W. Ryan (ed.), *Celiac Disease: Methods and Protocols*, Methods in Molecular Biology, vol. 1326,
DOI 10.1007/978-1-4939-2839-2_3, © Springer Science+Business Media New York 2015

1.1 Higher Order Models

Both the dog and monkey models are spontaneous models of gluten sensitivity [2, 3]. In the dog model, Irish Setters will develop intestinal permeability, partial villous atrophy, and increased numbers of intraepithelial lymphocytes (IELs) in response to gluten challenge [4]. Although this pathology is MHCII-independent, it is genetically transferable [5]. Irish Setters do not develop increased titers of antibodies against gliadin however, and so their utility may be restricted to the innate responses to gluten [6].

The nonhuman primate rhesus macaque model for celiac disease is also a spontaneous model, but requires screening for individuals that are gluten sensitive [3]. This process involves screening first for celiac related antibodies, then conducting biopsies of the small intestine to check for damage. Those monkeys that do develop disease similar to celiac disease exhibit gliadin associated intestinal permeability, IEL infiltration, and partial villous atrophy. It has also been observed in one study that a rhesus macaque developed dermatitis much like that of patients presenting with dermatitis herpetiformis. This monkey did not develop enteropathy, but did spontaneously generate antibodies against epidermal transglutaminase and tissue transglutaminase. These antibodies were decreased and the dermatitis resolved upon introduction of a gluten-free diet [7].

1.2 Rat Model

The germ-free Wistar AVN rat model is a nontransgenic, inducible model for studying celiac disease. Rats are kept germ free and administered gluten in increasing doses from birth for 2 months at which time they begin to exhibit villous atrophy, crypt hyperplasia, and increased numbers of IELs [8]. This model, paired with interferon-γ (IFN-γ) sensitization, has also been used to study the effects of gut bacteria in mediating gliadin-dependent changes in the intestinal epithelium [9, 10]. In one study, *Bifidobacterium bifidum*, *Shigella*, and *Escherichia coli* were surgically administered to separate sections of the rat intestine to determine the effects on epithelial permeability by the different bacteria [9]. *B. bifidum* protected against a gliadin or IFN-γ associated decrease in the tight junction protein ZO-1, as well as the translocation of gliadin beneath the epithelial layer. On the other hand, *E. coli* and *Shigella* were both associated with a decrease in the ZO-1 on the villi, as well as increased gliadin translocation across the epithelium. A proteome analysis showed a marked difference in the proteome pattern between rats fed gliadin with or without *B. bifidum*, supporting the idea that *B. bifidum* may have a therapeutic role for celiac disease [10].

1.3 Mouse Models

A spontaneous model for celiac disease has not yet been developed in the mouse, but many inducible models that represent gluten-driven inflammation have been generated. One inducible mouse model is the CD4$^+$ transfer model. In this model, CD4$^+$CD45RB^{low}CD25$^-$ T cells from gliadin sensitized B6 mice are transferred to the immunodeficient Rag1$^{-/-}$ mice (also on a B6 background) [11]. The recipient mice exhibited decreased weight gain as well as villous atrophy, crypt hyperplasia, and lymphocyte and neutrophil infiltration while on a gluten-containing diet, all of which resolved with the removal of gluten from the diet. Another inducible mouse model utilized Balb/c mice [12]. In this model, gluten-associated enteropathy was also induced by maintaining three successive generations of mice on a gluten-free diet, immunizing the fourth generation with whole gliadin emulsified in complete Freund's adjuvant (CFA), and then providing a gluten-containing standard diet.

1.4 Class II HLA Transgenic Murine Models

Because of the strong genetic component of celiac disease, there have been a number of transgenic mouse lines generated that contain the human MHCII genes, HLA-DQ2 and HLA-DQ8, which are strongly associated with celiac disease [13]. Indeed, celiac disease is the autoimmune disease that is most tightly associated with HLA molecules, with greater than 95 % of all celiac disease patients being HLA-DQ2 and HLA-DQ8. The remainder of this chapter will describe the generation of these mice and the advantages of the different mouse lines.

2 Methods Used to Generate Transgenic Models

2.1 Alternative Approaches

Over the past two decades, two different approaches have been taken to generate transgenic mice for celiac research. One approach is to insert genetic material that was derived from human patients, called genomic fragments; the second approach is to insert cDNA that was generated by removing the nucleotide sequences of the noncoding regions called introns (Fig. 1). Genomic fragments contain the original *human* promoters and regulatory elements associated with HLA-DQ2 or DQ8 genes, as well as any other genes that might be tightly linked (Fig. 1).

2.2 DQ8 Transgenic Constructs

Currently, there are at least three different DQ8 transgenic constructs that have been generated for use in mice [14–16]. All three lines were generated using genomic fragments that contain DQ8; however, they differed in the use of promoter elements to drive the expression. The line generated in Herman et al. used the I-Eα (mouse) promoter, whilst the other two lines used the human promoter elements within the genomic fragment transferred to the mice. The expression of DQ8 by different cell types has been well characterized in the transgenic mouse line that was generated by Dr. Chella David's group [15]. In this line of mice, DQ8 is expressed by classical antigen-presenting cells (B cells, macrophages, and dendritic cells) as well as non-classical antigen-presenting cells (epithelial cells) [17].

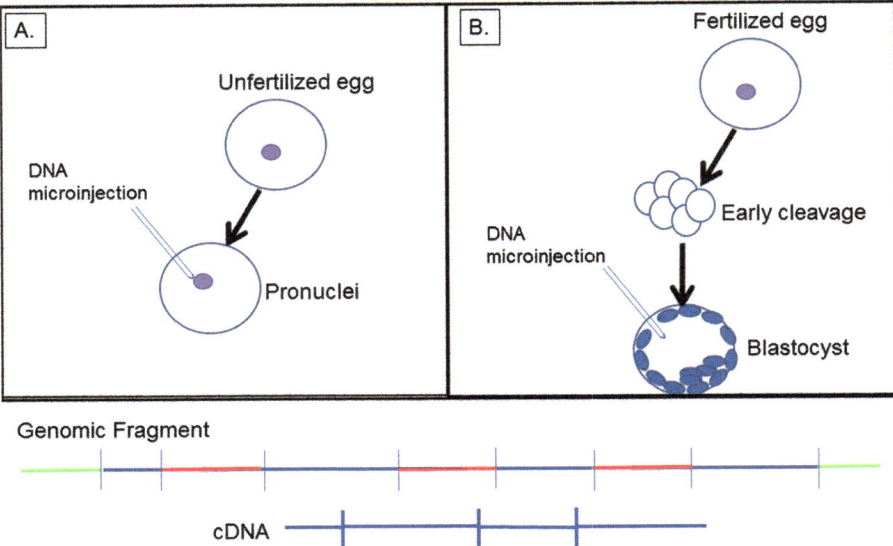

Fig. 1 Insertion of genetic constructs for generating hCD4 and HLA transgenic mice. Transgenic mice that express HLA molecules and/or human CD4 were generated by microinjecting DNA into either pronuclei (**a**) of an unfertilized egg or the cytoplasm of blastocysts (**b**). The DNA injected was either a genomic fragment or cDNA, in which the introns (*red*), promoter elements, and surrounding linked genes (*green*) have been removed, leaving only exons (*blue*)

2.3 DQ2 Transgenic Constructs

Similarly, three different lines of DQ2 transgenic mice have been generated using genomic fragments. In 2002, Chen et al. generated a mouse line containing a ~320 kb insertion of seven genes, spanning from HLA-DRA to HLA-DQB2, using a yeast artificial chromosome [18] (Fig. 2). This mouse was crossed with a partial murine MHCII knockout ($A\beta^{o/o}$). The resultant mouse line expressed HLA DR3 and DQ2 on dendritic cells, B cells, and macrophages, but not on T cells. While there was no intestinal pathology observed in these mice, they did generate a greater gliadin-specific T cell response than their non-DQ2 transgenic counterparts.

Another group also used yeast artificial chromosomes to incorporate a DQ2 containing genomic fragment into a mouse. The 550 kb fragment, which contained the complete TAP1/LMP, DRα, DRβ1, DRβ3, DQα, DQβ regions, was inserted into C57BL/6 mice, which were then crossed with an MHCII-IAβ⁰ mouse [19]. While no gluten-specific studies have been performed on these mice, ovalbumin (OVA) immunized mice did proliferate when presented OVA by human B-LCL cells in vitro, suggesting the capacity to mount an antigen-specific proliferative response. Of particular interest is that this transgenic mouse expressed low levels of DQ on resting and activated T cells (~12 % and 40 %, respectively). This is advantageous, since activated human T cells

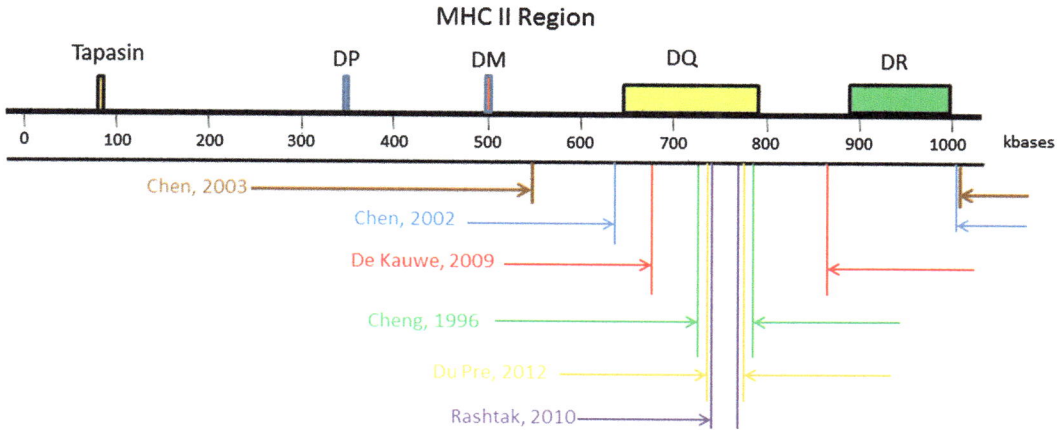

Fig. 2 The lengths of six different genetic constructs used to generate HLA transgenic mice. The size of each fragment is delineated on a cartoon of the MHCII genetic region by *two lines* on either side of the beginning and end of the fragment. The article that describes the genetic construct is listed to the *left* and is in the same color font as the lines depicting the size and location of the corresponding genetic fragment

express MHCII on their surfaces, and they may play a role in modulating disease in humans [20].

Yet another group, Du Pre et al., inserted a much smaller genomic fragment (68 kb) via microinjection [21]. This fragment contained DQA1*0501 and DQB1*0201 (DQ2.5), which is the suballele of DQ2 most tightly associated with celiac disease [22]. Additionally, these mice were crossed to gliadin-specific TCR transgenic mice to produce HLA-DQ2$^+$ gliadin-TCR.MHCII$^{\Delta/\Delta}$ double transgenic mice [21]. In the spleen DQ2 was expressed on B cells and DCs, while CD3$^+$ T cells expressed the transgenic TCR. CD4$^+$ T cells from the HLA-DQ2 gliadin-TCR.MHCII$^{\Delta/\Delta}$ double transgenic mouse were CFSE labeled and transferred intravenously into DQ2 mice to study where the gliadin-specific T cells are being activated and proliferating in vivo. Curiously, proliferating T cells could only be detected in the spleens of gliadin-fed mice, whereas in mice sensitized and challenged with OVA, proliferating T cells could be found in the Peyer's patches and mesenteric lymph nodes. The proliferating T cells expressed a pro-inflammatory phenotype, secreting high levels of IFN-γ, as well as IL-6 and IL-17. IL-10 was also detected, and its secretion was determined to be from two different subsets of T cells: one subset that secretes IL-10, but not IFN-γ, and another subset that expresses both IL-10 and IFN-γ. A third subset of proliferating splenic T cells secrete IFN-γ but not IL-10. However, as with the other DQ2 mouse lines, no enteropathy developed.

A different approach was taken by a fourth group, that of using DQ2 cDNA [23]. Because not all human promoter elements associated with the genomic fragment interface well with the mouse transcriptional components within the transgenic mouse, some experimental designs require that cDNA constructs of HLA with a known mouse derived promoter element be used. In doing so, transgene expression can be targeted to specific tissues, depending on what type of promoter is chosen. Rashtak et al. placed the H-2Eα promoter upstream of the cDNA sequences for both DQA1*05 and DQB1*02 [23]. The resultant transgenic mice were then crossed to MHCII$^{Δ/Δ}$ (AE^0) mice. The advantage to this mouse model is that the cDNA allows for expression of DQ2 without the expression of DR3 to better study the specific effects of DQ2 alone. These mice developed spontaneous skin lesions on their ears, muzzles, tails, and paws, epidermal thickening, and IgG and IgM deposits at the basement membrane, similar to what is described in patients with lupus, which is another autoimmune disease sometimes associated with celiac disease [24]. Thus, the development of a lupus-like disease in this DQ2 transgenic mouse demonstrates two points. The first is that expression of DQ2 alone (in the absence of DR3) does not result in celiac disease; the second is that expression of DQ2 alone can lead to the spontaneous development of an autoimmune disease associated with celiac disease.

2.4 Incorporation of Human CD4 Transgenes

In some of the lines of HLA-DQ2 and DQ8 transgenic mice discussed above, it was observed that there was a decreased number of CD4$^+$ T cells in their peripheral blood. In order to try to restore CD4$^+$ T cell numbers in these lines of HLA transgenic mice, the human CD4 transgene (hCD4) was incorporated into their genetic background. Currently there are at least two different transgenic constructs of hCD4 that were used to generate hCD4$^+$ transgenic mice. One construct was crossed with DQ8 mice to generate HCD4$^+$ DQ8$^+$ transgenic mice [25]. In this construct, human CD4 cDNA along with mouse CD2 promoter elements was used [26]. These mice had DQ8 restricted T cell responses to gluten after sensitization to gluten, but no villous atrophy developed. In later studies, these same mice were shown to have increased CD103$^+$ trans-epithelial cells after gliadin sensitization, as well as increased intestinal permeability and compromised tight junction structure [27, 28]. Treatment with larazotide acetate improved tight junction structure and decreased intestinal paracellular permeability [27], as well as treatment with a high dose of the intestinal paracellular inhibitor AT1001 [19], and treatment with the gliadin binding polymer P(HEMA-*co*-SS) [29]. Mice treated with P(HEMA-*co*-SS) before gliadin sensitization exhibited increased IL-10 and decreased TNF-α production in vitro. Particularly, P(HEMA-*co*-SS) seems to exert some therapeutic effects, as administration to HLA-HCD4/DQ8 mice restored villi to near nonsensitized levels [29].

2.5 NOD Genetic Background

In De Kauwe et al., hCD4$^+$ transgenic mice were crossed with DR3-DQ2.MHC$^{Δ/Δ}$ transgenic mice on C57Bl/6 and NOD backgrounds [30]. The hCD4$^+$ transgenic mice used were generated by using an hCD4 mini-gene, in which a large intron was removed [31]. Disruption of the mouse CD4 gene was achieved by a target insertion of a MC1neo cassette. The introduction of human CD4 increased CD4+ T cell numbers in these DQ2 transgenic mice, which may be the result of a more specific interaction between the HLA and human CD4. However, these mice did not spontaneously develop any gluten-related intestinal pathology.

Because celiac patients are at a higher risk for developing other autoimmune diseases, some groups incorporated the NOD genetic background, which is predisposed towards developing autoimmune diseases such as type 1 diabetes and thyroiditis [24, 32]. This genetic background is well defined and consists of over 27 Idd (Insulin Dependent Diabetes) loci, which contribute to the development of diabetes [33]. To address the effect of an autoimmune predisposing genetic background, NOD mice were crossed with DQ8 and DQ2 mice.

The first cross was between NOD mice and DQ8 transgenic mice, and the second was between DQ2DR3 mice and NOD mice [30, 34]. Neither of these mouse lines spontaneously developed gluten-dependent enteropathy. However, after sensitization to gluten, the NOD DQ8 mice did develop a gluten-dependent skin disease similar to dermatitis herpetiformis [34].

Because of this lack of spontaneous enteropathy in the HLA transgenic mice, it appears that other perturbations of the intestinal immune system of the HLA transgenic mouse lines must occur in order for this to develop. Indeed, many studies with the different lines of HLA transgenic mice would indicate this. Transient inductions of intestinal permeability such as the use of indomethacin have shown an increased number of intraepithelial cells in the HLA-DQ2 and DQ8 lines of mice [28, 35]. A study with NOD DQ8 mice found that partially depleting regulatory T cells (CD4+CD25+ Foxp3+ cells) with anti-CD25 antibody treatment for 2 weeks, followed by the administration of cholera toxin and gliadin together every week for 3 weeks, resulted in a mild increase in the number of CD3+ IELs (3 for every 100 enterocytes) over no treatment, and a decreased villous height/crypt ratio [36].

2.6 Villous Atrophy and Interleukin 15

Despite all of these variations of sensitization to gluten (or gluten-derived peptides) and simultaneous transient inflammation due to adjuvants, no villous atrophy characteristic of Marsh III has been observed to develop. Thus, chronic states of intestinal inflammation would appear to be required for the development of villous atrophy.

One example of chronic inflammation in celiac disease is the upregulation of IL-15 in the intestines of celiac patients [37]. IL-15 is a key inflammatory cytokine that has been shown to interfere with regulatory T cell development and function [38]. In 2002, a mouse line was generated that expressed human

IL-15 in enterocytes via the tissue-specific promoter T3b [39]. These mice developed spontaneous duodenal-jejunal inflammation, villous atrophy, crypt hyperplasia (height to crypt ratio, 2.1:1), as well as leukocyte infiltration of the lamina propria. However, while these symptoms are typical of patients with celiac disease, none of these manifestations were gluten specific in this IL-15 transgenic mouse line.

Another line of IL-15 transgenic mice uses a modified murine IL-15 transgene that lacks key posttranscriptional checkpoints to facilitate overexpression, and utilizes the MHC I D^d promoter for systemic expression [40]. These mice exhibit increased numbers of lymphocytes early on, particularly DX5+CD3− NK cells and CD8+CD44hiCD69−Ly6Chi T cells, and will spontaneously develop clonal lymphocytic leukemia at about 20 weeks. In 2011, DePaolo et al. used this mouse line to make an IL-15 HLA-DQ8 double transgenic mouse line, which generated gluten-specific IFN-γ producing CD4+ T cells in the lamina propria and mesenteric lymph nodes in mice on a gluten-containing diet [37]. Interestingly, this phenomenon was retinoic acid dependent, which also stimulated inflammatory T_H17 associated responses. These mice also presented elevated numbers of IELs; however, they did not display any villous atrophy, indicating this model is better suited to the study of the early stages of celiac disease over the later stages.

3   Conclusions

In summary then, HLA transgenic mice that express either DQ2 or DQ8 can be useful tools for evaluating the pathogenesis of celiac disease (Fig. 3), but a number of factors should be considered before starting. The first choice would be which HLA molecule is more appropriate for addressing the question of concern. The second decision should be on whether to use a cDNA construct or a genomic fragment. Which promoter elements are used to drive the expression of the HLA cDNA gene is equally important. Also to be considered is which suballele of DQ2 or DQ8 would be beneficial for the individual researcher's study, due to the difference in level of association of each suballele with celiac disease. In addition, the use of genetic manipulations such as overexpressing a transgene, knocking out genes, or targeted expression of genes for generating chronic inflammation is currently being incorporated into HLA transgenic mice. Finally, the genetic background should also be considered when developing a mouse model for celiac disease-related studies, as that would affect the development of autoimmune reactions and subsequent phenotypes.

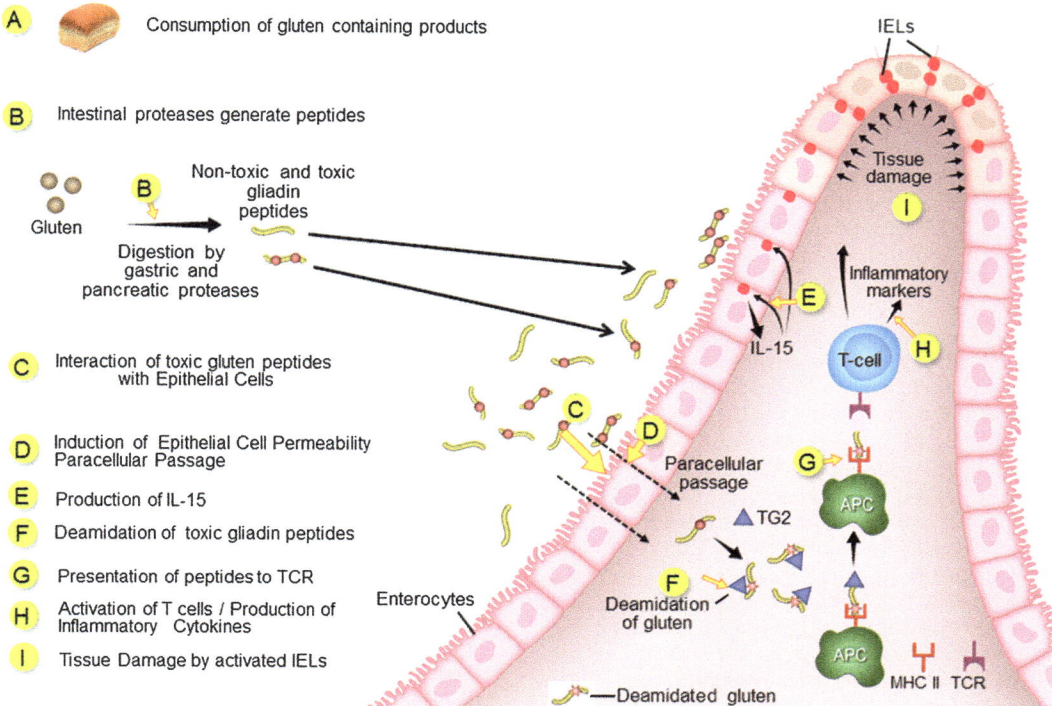

A: Consumption of gluten containing products

B: Intestinal proteases generate peptides

Gluten

Non-toxic and toxic gliadin peptides

Digestion by gastric and pancreatic proteases

C: Interaction of toxic gluten peptides with Epithelial Cells

D: Induction of Epithelial Cell Permeability Paracellular Passage

E: Production of IL-15

F: Deamidation of toxic gliadin peptides

G: Presentation of peptides to TCR

H: Activation of T cells / Production of Inflammatory Cytokines

I: Tissue Damage by activated IELs

IELs

Tissue damage

Inflammatory markers

IL-15

T-cell

Paracellular passage

TG2

APC

Enterocytes

Deamidation of gluten

APC

MHC II TCR

Deamidated gluten

Fig. 3 Pathogenesis of celiac disease. A number of events lead to the development of celiac disease in patients, not all of which have been shown to occur in a progressive timeline. *A*: Consumption of gluten-containing products. *B*: Intestinal proteases digest gluten into toxic and nontoxic peptides. *C*: Toxic gluten-derived peptides interact with epithelial cells. *D*: Toxic gluten peptides induce epithelial permeability, which then leads to the subsequent passage of gluten-derived peptides into the *lamina propria*. *E*: IL-15 is produced by epithelial cells. *F*: Gliadin-derived peptides are deamidated by tissue transglutaminase (tTG). *G*: Antigen-presenting cells then present deamidated gliadin-derived peptides via MHCII to T cells. *H*: Subsequent activation of T cells occurs and inflammatory cytokines are produced. *I*: Tissue damage is mediated by activated intraepithelial lymphocytes (IELs)

References

1. van der Kolk JH, van Putten LA, Mulder CJ, Grinwis GC, Reijm M, Butler CM, von Blomberg BM (2012) Gluten-dependent antibodies in horses with inflammatory small bowel disease (ISBD). Vet Q 32:3–11

2. Batt RM, Carter MW, McLean L (1984) Morphological and biochemical studies of a naturally occurring enteropathy in the Irish setter dog: a comparison with coeliac disease in man. Res Vet Sci 37:339–346

3. Bethune MT, Borda JT, Ribka E, Liu MX, Phillippi-Falkenstein K, Jandacek RJ, Doxiadis GG, Gray GM, Khosla C, Sestak K (2008) A non-human primate model for gluten sensitivity. PLoS One 3:e1614

4. Hall EJ, Batt RM (1992) Dietary modulation of gluten sensitivity in a naturally occurring enteropathy of Irish setter dogs. Gut 33:198–205

5. Polvi A, Garden OA, Houlston RS, Maki M, Batt RM, Partanen J (1998) Genetic susceptibility to gluten sensitive enteropathy in Irish setter dogs is not linked to the major histocompatibility complex. Tissue Antigens 52: 543–549

6. Hall EJ, Carter SD, Barnes A, Batt RM (1992) Immune responses to dietary antigens in gluten-sensitive enteropathy of Irish setters. Res Vet Sci 53:293–299

7. Sestak K, Mazumdar K, Midkiff CC, Dufour J, Borda JT, Alvarez X (2011) Recognition of

epidermal transglutaminase by IgA and tissue transglutaminase 2 antibodies in a rare case of Rhesus dermatitis. J Vis Exp PMCID: 3369644

8. Tlaskalova-Hogenova H, Stepankova R, Farre M, Funda DP, Rehakova Z, Sinkora J, Tuckova L, Horak I, Horakova D, Cukrowska B, Kozakova H, Kolinska J (1997) Autoimmune reactions induced by gliadin feeding in germ-free AVN rats and athymic nude mice. Animal models for celiac disease. Ann N Y Acad Sci 815:503–505

9. Cinova J, De Palma G, Stepankova R, Kofronova O, Kverka M, Sanz Y, Tuckova L (2011) Role of intestinal bacteria in gliadin-induced changes in intestinal mucosa: study in germ-free rats. PLoS One 6:e16169

10. Olivares M, Laparra M, Sanz Y (2012) Oral administration of Bifidobacterium longum CECT 7347 modulates jejunal proteome in an in vivo gliadin-induced enteropathy animal model. J Proteomics 77:310–320

11. Freitag TL, Rietdijk S, Junker Y, Popov Y, Bhan AK, Kelly CP, Terhorst C, Schuppan D (2009) Gliadin-primed CD4+CD45RBlowCD25− T cells drive gluten-dependent small intestinal damage after adoptive transfer into lymphopenic mice. Gut 58:1597–1605

12. Papista C, Gerakopoulos V, Kourelis A, Sounidaki M, Kontana A, Berthelot L, Moura IC, Monteiro RC, Yiangou M (2012) Gluten induces coeliac-like disease in sensitised mice involving IgA, CD71 and transglutaminase 2 interactions that are prevented by probiotics. Lab Invest 92:625–635

13. Stepniak D, Koning F (2006) Celiac disease—sandwiched between innate and adaptive immunity. Hum Immunol 67:460–468

14. Wen L, Wong FS, Burkly L, Altieri M, Mamalaki C, Kioussis D, Flavell RA, Sherwin RS (1998) Induction of insulitis by glutamic acid decarboxylase peptide-specific and HLA-DQ8-restricted CD4(+) T cells from human DQ transgenic mice. J Clin Invest 102:947–957

15. Cheng S, Baisch J, Krco C, Savarirayan S, Hanson J, Hodgson K, Smart M, David C (1996) Expression and function of HLA-DQ8 (DQA1*0301/DQB1*0302) genes in transgenic mice. Eur J Immunogenet 23:15–20

16. Herman AE, Tisch RM, Patel SD, Parry SL, Olson J, Noble JA, Cope AP, Cox B, Congia M, McDevitt HO (1999) Determination of glutamic acid decarboxylase 65 peptides presented by the type I diabetes-associated HLA-DQ8 class II molecule identifies an immunogenic peptide motif. J Immunol 163:6275–6282

17. Chapoval SP, Nabozny GH, Marietta EV, Raymond EL, Krco CJ, Andrews AG, David CS (1999) Short ragweed allergen induces eosinophilic lung disease in HLA-DQ transgenic mice. J Clin Invest 103:1707–1717

18. Chen Z, Dudek N, Wijburg O, Strugnell R, Brown L, Deliyannis G, Jackson D, Koentgen F, Gordon T, McCluskey J (2002) A 320-kilobase artificial chromosome encoding the human HLA DR3-DQ2 MHC haplotype confers HLA restriction in transgenic mice. J Immunol 168:3050–3056

19. Chen D, Ueda R, Harding F, Patil N, Mao Y, Kurahara C, Platenburg G, Huang M (2003) Characterization of HLA DR3/DQ2 transgenic mice: a potential humanized animal model for autoimmune disease studies. Eur J Immunol 33:172–182

20. Mangalam A, Rodriguez M, David C (2006) Role of MHC class II expressing CD4+ T cells in proteolipid protein(91-110)-induced EAE in HLA-DR3 transgenic mice. Eur J Immunol 36:3356–3370

21. Du Pre MF, Kozijn AE, van Berkel LA, ter Borg MN, Lindenbergh-Kortleve D, Jensen LT, Kooy-Winkelaar Y, Koning F, Boon L, Nieuwenhuis EE, Sollid LM, Fugger L, Samsom JN (2011) Tolerance to ingested deamidated gliadin in mice is maintained by splenic, type 1 regulatory T cells. Gastroenterology 141:610–620, 620.e1–2

22. Fallang LE, Bergseng E, Hotta K, Berg-Larsen A, Kim CY, Sollid LM (2009) Differences in the risk of celiac disease associated with HLA-DQ2.5 or HLA-DQ2.2 are related to sustained gluten antigen presentation. Nat Immunol 10:1096–1101

23. Rashtak S, Marietta E, Cheng S, Camilleri M, Pittelkow M, David C, Grande J, Murray J (2010) Spontaneous lupus-like syndrome in HLA-DQ2 transgenic mice with a mixed genetic background. Lupus 19:815–829

24. Rashtak S, Marietta EV, Murray JA (2009) Celiac sprue: a unique autoimmune disorder. Expert Rev Clin Immunol 5:593–604

25. Black KE, Murray JA, David CS (2002) HLA-DQ determines the response to exogenous wheat proteins: a model of gluten sensitivity in transgenic knockout mice. J Immunol 169:5595–5600

26. Paterson RK, Burkly LC, Kurahara DK, Dunlap A, Flavell RA, Finkel TH (1994) Thymic development in human CD4 transgenic mice. Positive selection occurs after commitment to the CD8 lineage. J Immunol 153:3491–3503

27. Gopalakrishnan S, Durai M, Kitchens K, Tamiz AP, Somerville R, Ginski M, Paterson BM, Murray JA, Verdu EF, Alkan SS, Pandey NB (2012) Larazotide acetate regulates epithelial

tight junctions in vitro and in vivo. Peptides 35:86–94

28. Silva MA, Jury J, Sanz Y, Wiepjes M, Huang X, Murray JA, David CS, Fasano A, Verdu EF (2012) Increased bacterial translocation in gluten-sensitive mice is independent of small intestinal paracellular permeability defect. Dig Dis Sci 57:38–47

29. Pinier M, Fuhrmann G, Galipeau HJ, Rivard N, Murray JA, David CS, Drasarova H, Tuckova L, Leroux JC, Verdu EF (2012) The copolymer P(HEMA-co-SS) binds gluten and reduces immune response in gluten-sensitized mice and human tissues. Gastroenterology 142:316–325.e1–12

30. de Kauwe AL, Chen Z, Anderson RP, Keech CL, Price JD, Wijburg O, Jackson DC, Ladhams J, Allison J, McCluskey J (2009) Resistance to celiac disease in humanized HLA-DR3-DQ2-transgenic mice expressing specific anti-gliadin CD4+ T cells. J Immunol 182:7440–7450

31. Killeen N, Davis CB, Chu K, Crooks ME, Sawada S, Scarborough JD, Boyd KA, Stuart SG, Xu H, Littman DR (1993) CD4 function in thymocyte differentiation and T cell activation. Philos Trans R Soc Lond B Biol Sci 342:25–34

32. Ellis JS, Wan X, Braley-Mullen H (2013) Transient depletion of CD4+ CD25+ regulatory T cells results in multiple autoimmune diseases in wild-type and B-cell-deficient NOD mice. Immunology 139:179–186

33. Driver JP, Serreze DV, Chen YG (2011) Mouse models for the study of autoimmune type 1 diabetes: a NOD to similarities and differences to human disease. Semin Immunopathol 33:67–87

34. Marietta E, Black K, Camilleri M, Krause P, Rogers RS III, David C, Pittelkow MR, Murray JA (2004) A new model for dermatitis herpetiformis that uses HLA-DQ8 transgenic NOD mice. J Clin Invest 114:1090–1097

35. Natividad JM, Huang X, Slack E, Jury J, Sanz Y, David C, Denou E, Yang P, Murray J, McCoy KD, Verdu EF (2009) Host responses to intestinal microbial antigens in gluten-sensitive mice. PLoS One 4:e6472

36. Galipeau HJ, Rulli NE, Jury J, Huang X, Araya R, Murray JA, David CS, Chirdo FG, McCoy KD, Verdu EF (2011) Sensitization to gliadin induces moderate enteropathy and insulitis in nonobese diabetic-DQ8 mice. J Immunol 187:4338–4346

37. DePaolo RW, Abadie V, Tang F, Fehlner-Peach H, Hall JA, Wang W, Marietta EV, Kasarda DD, Waldmann TA, Murray JA, Semrad C, Kupfer SS, Belkaid Y, Guandalini S, Jabri B (2011) Co-adjuvant effects of retinoic acid and IL-15 induce inflammatory immunity to dietary antigens. Nature 471:220–224

38. Zanzi D, Stefanile R, Santagata S, Iaffaldano L, Iaquinto G, Giardullo N, Lania G, Vigliano I, Vera AR, Ferrara K, Auricchio S, Troncone R, Mazzarella G (2011) IL-15 interferes with suppressive activity of intestinal regulatory T cells expanded in Celiac disease. Am J Gastroenterol 106:1308–1317

39. Ohta N, Hiroi T, Kweon MN, Kinoshita N, Jang MH, Mashimo T, Miyazaki J, Kiyono H (2002) IL-15-dependent activation-induced cell death-resistant Th1 type CD8 alpha beta+NK1.1+ T cells for the development of small intestinal inflammation. J Immunol 169:460–468

40. Fehniger TA, Suzuki K, Ponnappan A, VanDeusen JB, Cooper MA, Florea SM, Freud AG, Robinson ML, Durbin J, Caligiuri MA (2001) Fatal leukemia in interleukin 15 transgenic mice follows early expansions in natural killer and memory phenotype CD8+ T cells. J Exp Med 193:219–231

41. Catassi C, Fabiani E, Rätsch IM et al (1996) The coeliac iceberg in Italy. A multicentre anti-gliadin antibodies screening for coeliac disease in school-age subjects. Acta Paediatr 85:29–35. doi:10.1111/j.1651-2227.1996.tb14244.x

Chapter 4

Study Designs for Exploring the Non-HLA Genetics in Celiac Disease

Åsa Torinsson Naluai

Abstract

Since the breakthrough of genome-wide association studies and genetic studies of common complex diseases like celiac disease have been able to finally identify reproducible gene regions affecting risk of developing disease. Before it was possible to perform genome-wide association analysis, the field struggled with genome-wide linkage analysis to identify gene regions. Genome-wide linkage had been very successful in identifying genes underlying monogenic diseases, but common complex polygenic diseases did not prove so tractable.

This chapter will describe the genome-wide methods available for genetic analyses of families today and compare these with the previous analyses performed in the 1990s and early twenty-first century.

Key words Genetic linkage, Association, Complex inheritance, Genome-wide, Sib-pairs, Transmission disequilibrium test, Nonparametric linkage, LOD score

1 Introduction

The causes of susceptibility to common diseases are still poorly understood. The molecular mechanisms involved in initiating celiac disease are, as in most diseases, largely unknown. With genetic analyses we can identify which genes are involved and start to understand the molecular mechanisms responsible for disease onset. Most importantly, this can be done without *previous conceptions* as to what these mechanisms are. In this way, whole genome analysis is completely hypothesis free and not limited to the knowledge of today. Instead it will generate new hypotheses, which can then be investigated further.

Our genes influence all of our diseases in one way or another and almost all human traits and diseases show interindividual differences in a population. Even though environmental factors are believed to play a role in most of our common diseases, human genetic variation is a major contributor to differences in susceptibility to disease and is also part of the reason why some diseases run

Anthony W. Ryan (ed.), *Celiac Disease: Methods and Protocols*, Methods in Molecular Biology, vol. 1326,
DOI 10.1007/978-1-4939-2839-2_4, © Springer Science+Business Media New York 2015

in families. When it comes to polygenic diseases, there is no simple answer for how to distinguish disease influencing genetic variants from nonfunctional genetic variants. Often, the "disease" variants are present in healthy individuals or even very common in healthy individuals (such as the HLA variants in celiac disease and other autoimmune diseases)

In this chapter we will discuss how we can identify the genetic variation that influences disease susceptibility using families with affected and unaffected individuals. We will focus on whole genome genetic mapping methods for common complex diseases like celiac disease and to a lesser extent on mapping genes responsible for monogenic diseases. The methods described include linkage analysis, association analysis, and the transmission disequilibrium test (the TDT). Furthermore, we will present some of the available pathway analysis tools and ways to use existing public databases to "get more out of your data."

2   Methods

2.1   Linkage Analysis

The concept of genes being linked together and either transmitted to offspring together or separated by the so-called recombination events was known already long before the structure of DNA was known. We now know that the DNA strand comes in a set number of pieces per species (chromosomes) and that genes are located along these chromosomes. Each person has two copies of each chromosome; one is inherited from their mother and one from their father. When passing a gene onto the next generation, there is a 50 % chance that a particular gene will be transmitted from the maternal chromosome and a 50 % chance that it will be transmitted from the paternal chromosome. Since the chromosomes are "cut" up in smaller pieces due to recombination events during meiosis, even genes that are located on the same chromosome will not always be inherited together. Some genes from each "new" chromosome in each egg or sperm will be derived from the mother and some from the father. In this way, genes located on the same chromosome will then be transmitted to the next generation either together from one ancestral chromosome or separated by a recombination event and transmitted to the offspring from different ancestral chromosomes. The closer two genes are on the chromosome the more likely they are to be transmitted together to the next generation without being separated by a recombination event or "cross-over." If two genes are transmitted more often together than 50 % we say that they are "linked" and hence the name "linkage analysis" and Linkage Disequilibrium (LD) as opposed to Linkage Equilibrium. Genes located far away from each other on the chromosome (or on completely different chromosomes) are likely to be separated and you would then transmit one of these genes originating from the father and one from the mother, onto the offspring. To perform a linkage analysis, the transmission of genetic variation between generations in families is tracked.

Genetic mapping using linkage analysis started already in the early 1900s; however, there were not enough genetic markers identified that could be analyzed and tracked from generation to generation. The first paper suggesting this methodology for scanning all genes in the whole genome was written by David Botstein and colleagues and published in Am J Hum Genet in 1980. Shortly after, in 1983, Kary Mullis invented the Polymerase Chain Reaction (PCR) technology. These events led to an incredible boost in linkage analysis, and genes for over 4000 monogenic diseases have been identified this way since the 1980s and 1990s, the first one being the gene for Huntington's disease in 1983 (OMIM database, #613004).

After seeing the success in rare monogenic diseases, attention turned to linkage analyses to find the genetic background for common diseases with a complex polygenic inheritance pattern. Several genome-wide linkage scans for celiac disease were performed, the first one by Greco and coworkers in 1998 [1] and several more were to follow including our own in 2001 [2]. Although no specific gene was found, several studies were able to replicate the linkage peak on chromosome 5, discovered by the first genome scan in celiac disease [3].

For complex diseases like celiac disease it has been common to use the so-called nonparametric linkage analyses (NPL). However, LOD-score methods can also be used. The difference is that for LOD-score methods, certain parameters that describe the expected disease alleles and the inheritance pattern for these must be set. These parameters involve the allele frequency of the disease variant, and whether it is inherited in a dominant, recessive, or sex linked manner. Since there are many disease variants in a complex disease, different parameters must be tested to see which ones generate interesting results. For NPL analysis, it is not necessary to set any parameters, since this method simply counts the number of times a certain region has been inherited identical by descent (IBD) between two affected siblings. If two siblings have inherited a region IBD, they have inherited the same parental chromosome from for example their father. If they inherit the same parental chromosome from both their father and their mother, they have inherited two chromosomes IBD (2 IBD) in that particular region surrounding the marker. The more sibling pairs that inherit 2 IBD for a certain marker, the higher linkage score for that marker. There are many different programs, which can run family-based linkage analysis.

Two examples of recent software packages are Merlin [4] and Allegro [5]. Allegro can be described as a faster version of GENEHUNTER [6] (30-fold increased speed) and is freely available from DeCode genetics http://www.decode.com/software/. Allegro can handle somewhat larger pedigrees and computes both nonparametric allele sharing (NPL) and parametric multipoint LOD scores. Merlin presently does not compute parametric multipoint scores. Merlin is also freely available, (http://www.sph.umich.edu/csg/abecasis/Merlin) and carries out singlepoint and multipoint analyses in families. This includes IBD estimations and nonparametric linkage analysis for traits with affected verses non-affected individuals, as well as variance component linkage analysis, which is used for quantitative traits loci mapping (QTL). Merlin also has a function to limit the number of recombinants between consecutive markers and estimate haplotypes. It can list all possible nonrecombinant haplotypes within short regions. An appealing feature in Merlin is that it can identify genotyping errors that would be missed by most other programs. This is especially important when using SNPs, for which genotyping errors will less often lead to Mendelian inconsistencies than microsatellite markers. Merlin provides support for handling very large numbers of markers as well as gene-dropping simulations for estimating empirical significance levels.

Today, SNP arrays are just as well used for linkage as for association analysis. A typical highly informative SNP array is very useful for linkage analysis and is much easier to analyze compared with microsatellites. However, from the total set of SNP markers, an LD pruned set should be selected. Parameters of a window size of 50 and $R^2 < 0.5$ are recommended. This removes all uninformative markers and all markers in strong LD, and this can be achieved using the freely available software package PLINK [7]. After pruning and replacing all Mendel errors with "0," one can export a ped-file and a map-file and run Linkage analysis on the software package of choice. (The Decode genetic map is automatically supplied with the marker information from Illumina.) However, when using PLINK for replacing Mendel errors, one should be aware that PLINK only detects simple errors in small nuclear families. For a more thorough Mendelian check, the function in the Merlin software package may be used.

We have run the same families using both the new SNP array technology and the "old" microsatellite technology. Apart from the difference in time that it took to run the samples and get the genotypes out (2 weeks compared with 2 years in our case), the information content was nearly 100 % with the SNP arrays and for the microsatellites it averaged around 50–80 % at the very best. This increased information content led to 13 identified regions with an NPL point wise p-value below 0.01 compared with 1 single region in the microsatellite scan [8].

2.2 Genome-Wide Association Studies (GWAS)

The year 2007 saw a breakthrough for Genome-Wide Association Studies (GWAS) [9]. More genetic risk factors for common diseases were identified this year, than had been previously reported overall.

This new methodology is used to analyze several hundreds of thousands or millions of genetic variants across the whole genome, each consisting of small changes in the DNA sequence, the so-called single nucleotide polymorphisms (SNPs) or copy number variants (CNVs).

In spite of the success of identifying new risk genes, the first generation of GWASs only revealed a small percentage of the genetic variance responsible for disease. The vast majority of the genetic factors are still to be found, commonly referred to as "the missing heritability" [10]. Possible disease variation including rare variants with a larger effect could not be detected.

2.3 Possible New Approaches to GWAS Study Design

The field of genomics and in particular GWAS has been accused of "little useful result per dollar input." In their essay in Cell 2010, McClellan and King wrote, "This degree of allelic, locus, and phenotypic heterogeneity has important implications for gene discovery. In particular, causality in this context can almost never be resolved by large-scale association or case-control studies" [11]. Instead, they suggest the use of next generation sequencing (NGS) techniques to uncover rare variants behind common diseases. However, the use of NGS is still expensive and time consuming. There are a few ways to increase the chance of detecting rare variants and to hopefully reveal some of the missing heritability without the involvement of the NGS technique. These strategies have great potential and have not been fully explored in most diseases.

For example (1) to take into account already known genetic risk factors, like the HLA locus; (2) to stratify the patient material in some way which selects for individuals that carry a rare variant, i.e., by subphenotype or by using an isolated ethnic population; (3) to use families and combine both genetic linkage and association; and finally (4) to use gene-gene interaction models and pathway analyses to extract more information from the data, i.e., using GWAS results below the genome-wide significant threshold.

We have taken advantage of some of these strategies to reveal part of the missing heritability in celiac disease in our recently published Linkage GWAS [8].

2.4 Using Families to Combine Genome-Wide Linkage and Genome-Wide Association

In diseases with a complex inheritance pattern where many genetic and environmental risk factors play a role, one would like to eliminate environmental influences in order to more easily identify the genetic risk factors.

When it comes to the statistical power of case-control studies versus family studies, it is not possible to simply compare these two designs and determine which one is more powerful. Many factors come into play like the frequency and number of disease variants, the population under study, and so on. Under some circumstances a case-control design can be more powerful and under other circumstances a family design is more powerful. One advantage with case-control designs it that it is possible to select an enhanced set of matched controls. This can mean for example very old controls without dementia compared with a young cohort of dementia patients, or heavy smokers without lung cancer compared with lung cancer patients with or without a smoking history, or lean type 2 diabetes patients versus obese controls. However, in some situations, a family material can be a very good complement to a case-control design, and in many situations even a better choice. Families with many members affected with a disease are more likely to carry more genetic risk factors compared with sporadic cases. Familial cases tend to be enriched for disease-predisposing alleles and there is an increased power especially for detecting rare genetic variants [12].

Already in 1996, Risch and Merikangas suggested to run Genome-Wide Association Studies (GWAS) on sib-pairs with the powerful TDT [13]. The TDT, introduced by Spielman et al. in 1993 [14], is implemented in PLINK [7]. Families with several affected children can be used to detect linkage in the presence of association using these highly informative SNP arrays. There is no bias to use families in this manner, since the analysis is simply a linkage analysis in the presence of association. However, if linkage to a certain region has previously been found in the study families, it is not possible to use the same families again and apply the TDT as a test of confirmation.

Risch and Merikangas also estimated that only between 50 and 260 affected sib-pairs are needed to find a disease variant of 10 % frequency in the population and if the genotypic risk ratio is somewhere between two times or four times the risk compared with not having the risk genotype. If there is only one case per family, 150–700 cases would be needed to reach the same power for that particular disease gene frequency [13].

Being able to incorporate linkage information leads to another advantage of using families for association studies, namely the robustness against population differences (population stratification). This means that different ethnic populations differ in their genetic makeup. It can become a serious problem if the cases and controls have somewhat different ancestry not related to the disease. Even within small regions on the map, there can be substantial differences in genetic background.

In family samples it is also possible to use an additional quality control step involving the Mendelian laws of inheritance. Offspring must have alleles from both parents. For example, if a parent is AA homozygote and the other parent is GG homozygote, the children must all be AG heterozygote. If this is not the case we will get a Mendelian inconsistency or a Mendelian inheritance "error." PLINK [7] can also be used to detect these Mendelian errors. If there were any Mendelian inconsistencies for a specific family, all subjects in the family should be set as a "missed call" or removed for that certain SNP.

2.5 Using Known Risk Genes in the Analysis to Find New Genes

We have used a stratified TDT analysis where trios were split into a low-risk and a high-risk group based on the HLA genotype of the affected offspring. Children carrying the HLA-DQA1*02/05 risk allele and homozygous for the HLA-DQB1*02 risk allele (i.e., individuals carrying the DR3/DR3 or the DR3/DR7 haplotypes) were put in the "high-risk" group and the remaining children were put in the "low-risk" group. The rationale behind this is comparable to dividing never-smokers and smokers in separate groups in a study on lung cancer. Individuals, who develop lung cancer and do not smoke, could be expected to have different risk factors or a higher dose of the genetic risk factors compared with the smoking group. Similarly, to increase the chances of finding associated genes, one can use only HLA positive controls in a case-control study of an autoimmune HLA associated disease or only smoking controls in a study on lung cancer. In a multifactorial disease setting, individuals who are healthy "in spite" of an already known risk factor are even more likely to not carry additional risk factors than for example "perfectly matched" controls or population controls.

2.6 Pathway Analysis

Since many markers just below genome-wide significance are still expected to be true findings, it is desirable to try and separate these from the true negative findings (those that show linkage and association close to genome-wide significance just by chance). In order to do so, pathway analysis as well as for example a two-locus interaction test may be used. We have analyzed connections between genes in different regions, using DAVID Bioinformatics database (http://david.abcc.ncifcrf.gov/), GeneTrail [15], and the Ingenuity Pathway Analysis (IPA) software (Ingenuity Inc. CA, USA). To define regions below genome-wide significance and to get a reasonable number of genes for the pathway and interaction analyses, somewhere between 200 and 500 regions would be usable. In our GWAS, we defined 383 regions using the following inclusion criteria: (1) p-value less than 3.0×10^{-4}; (2) p-value less than 0.01 in our analysis and with a p-value less than 0.05 in the GWAS by Dubois et al. [16] and if the product of these p-values were less than 5.0×10^{-5} and the association were in the same allelic direction; and (3) an allele transmission ratio of <0.2 or >5 combined with a p-value less than 2×10^{-3}. However, these cutoffs are somewhat arbitrary and could, for example, have been set to the top 1000 SNP associations.

With regions defined, one can also analyze all pairwise interactions using one SNP per region and a Likelihood Ratio (LR) test [8].

Because many available pathway analysis tools are developed for gene expression arrays, they do not usually take into account that genes surrounding associated SNPs sometimes all come from the same gene family. For example, in our celiac GWAS, surrounding the same SNP could be 10 Interleukin genes or 20 genes encoding histone proteins. Depending on how many gene clusters like this are present in the genome, these functions can be highly significant in the pathway analysis even if they only come from one single hit (SNP). We tried to solve this issue by removing all but one gene from the same gene family within each associated region. We simply classified genes as belonging to the same gene family when their names started with the same three letters.

The Broad Institute at MIT and Harvard have produced numerous software tools for genomic analyses (http://www.broadinstitute.org/scientific-community/software). We have used a few of these in our studies, such as "GRAIL" [17] and "SNAP." We used GRAIL as an easy tool to define the genes surrounding our GWAS SNPs. GRAIL uses known recombination hotspots in order to limit the region of interest surrounding each SNP marker. After gene regions are defined by GRAIL, they can be exported and manually curated for gene families. The SNPs, which are not available in GRAIL, can be submitted into the "Table" function in the UCSC Genome Browser (http://genome.ucsc.edu) database. In this way one can extract all genes or the closest genes from a certain distance (250,000 kb) surrounding our SNPs. SNAP is a useful tool to draw figures of the associated SNPs and their LD pattern with surrounding SNPs at the same time as it displays the p-values. SNAP is also used to find SNP markers, which are in Linkage Disequilibrium (LD), and to select SNPs in absolute LD that can serve as "proxies" for each other.

2.7 SNP Arrays (SNP Chip)

There are several different SNP arrays on the market. However, there are only two major companies manufacturing these SNP arrays, namely Illumina and Affymetrix. The former of the two have taken quite a large proportion of the sales regarding SNP arrays and they have also recently launched a focused array for a reasonable price (the human core array). This array can be combined with their exome array and is then called the "core exome" array. Both Affymetrix and Illumina also provide the possibility to add your own custom SNPs to an existing array. For example you can choose 4000 SNPs from the GWAS catalog of already associated SNPs for many diseases and traits or if you have a special interest in certain regions or genes from previous linkage analyses in your own disease.

When you run the arrays the samples are usually prepared in batches of 96 samples. Since there can be batch differences between experimental runs, you should make sure that all individuals in the same family are located on the same 96-well plate. Quality control should be done using a somewhat stringent call rate and in our study we choose to exclude all markers not reaching a 97 % call rate. However, if your DNA samples have good quality, most markers will easily pass this threshold.

2.8 Linkage Analysis Versus Association Analysis

By performing a pure linkage test, you will add some information about disease loci to the information gained by association analysis. For example, it is very important to remember that in those cases where you have many rare variants within the same gene, pure linkage will pick this up and an association analysis will most likely not. There can also be regions where several genes located close to each other are all contributing to the disease susceptibility. This is also not likely to be picked up by the association analysis. In celiac disease, chromosome 5 will probably include susceptibility variants of both or one of these two scenarios. Such regions could be especially interesting to target for Next Generation Sequencing (NGS technique). Using NGS for whole genome sequencing for a common complex disease will pose a number of problems, which are beyond the scope of this chapter.

Altogether, there are probably over a 1000 genes fine-tuning an individual's risk for disease and so far we have only touched on some of the low hanging fruit. Even if a reliable genetic risk profile to predict disease could not be identified, the results from both linkage and association studies would point towards certain genes and certain molecular signaling pathways as the most probable. This could help demonstrate mechanisms behind autoimmunity on a molecular level and provide novel targets and biomarkers for diagnosing and treating disease. It will also most likely to be possible in the near future to combine the genetic information available to confidently detect individuals at a very high risk, before they become affected with a potentially irreversible autoimmune disease.

2.9 Pediatric Versus Late Onset Celiac Disease

Just as it makes sense to stratify family material in cases with different known risk factors like double or single copies of the HLA risk alleles and analyze the genetic background of these groups separately, it can make sense to stratify cases in different phenotypically distinct groups. In celiac disease, one can become affected very early in life or later in life, and one such division of groups is to use this age of onset criteria. Children who get their diagnosis early could possibly have a distinct genetic makeup compared with those who get their disease late in life or as adults. A few studies have looked at genetic factors in relation to age of onset in celiac disease, with somewhat conflicting results [18–20]. Now that so much more data from GWAS studies have become available, much more can be done in this area.

3 Conclusion

Family studies contribute considerably to our understanding of complex common diseases such as celiac disease. In the last few years the focus has been on GWAS studies using case-control materials. However, now is the time to bring out the valuable family samples from the previous linkage era and to analyze these samples, using new technology and whole genome SNP arrays, for linkage analysis in the presence of association.

4   URLs

PLINK (http://pngu.mgh.harvard.edu/purcell/plink/)

GWAS catalog, http://www.genome.gov/gwastudies/

GRAIL (http://www.broadinstitute.org/mpg/grail/)

SNAP (http://www.broadinstitute.org/mpg/snap/ldplot.php/)

GeneTrail (http://genetrail.bioinf.uni-sb.de/)

DAVID http://david.abcc.ncifcrf.gov/

KEGG (www.genome.jp/kegg/)

Gene Ontology (www.geneontology.org/)

References

1. Greco L et al (1998) Genome search in celiac disease. Am J Hum Genet 62(3):669–675

2. Naluai AT et al (2001) Genome-wide linkage analysis of Scandinavian affected sib-pairs supports presence of susceptibility loci for celiac disease on chromosomes 5 and 11. Eur J Hum Genet 9(12):938–944

3. Babron MC et al (2003) Meta and pooled analysis of European coeliac disease data. Eur J Hum Genet 11(11):828–834

4. Cook EH Jr (2002) Merlin: faster linkage analysis with improved genotyping error detection. Pharmacogenomics J 2(3):139–140

5. Gudbjartsson DF et al (2000) Allegro, a new computer program for multipoint linkage analysis. Nat Genet 25(1):12–13

6. Kruglyak L et al (1996) Parametric and nonparametric linkage analysis: a unified multipoint approach. Am J Hum Genet 58(6):1347–1363

7. Purcell S et al (2007) PLINK: a tool set for whole-genome association and population-based linkage analyses. Am J Hum Genet 81(3):559–575

8. Östensson M et al (2013) A possible mechanism behind autoimmune disorders discovered by genome-wide linkage and association analysis in celiac disease. PLoS One 8(8):e70174

9. Pennisi E (2007) Breakthrough of the year. Human genetic variation. Science 318(5858):1842–1843

10. Maher B (2008) Personal genomes: the case of the missing heritability. Nature 456(7218):18–21

11. McClellan J, King MC (2010) Genetic heterogeneity in human disease. Cell 141(2):210–217

12. Risch N, Teng J (1998) The relative power of family-based and case-control designs for linkage disequilibrium studies of complex human diseases I. DNA pooling. Genome Res 8(12):1273–1288

13. Risch N, Merikangas K (1996) The future of genetic studies of complex human diseases. Science 273(5281):1516–1517

14. Spielman RS, McGinnis RE, Ewens WJ (1993) Transmission test for linkage disequilibrium: the insulin gene region and insulin-dependent diabetes mellitus (IDDM). Am J Hum Genet 52(3):506–516

15. Backes C et al (2007) GeneTrail—advanced gene set enrichment analysis. Nucleic Acids Res 35(Web Server issue):W186–W192

16. Dubois PC et al (2010) Multiple common variants for celiac disease influencing immune gene expression. Nat Genet 42(4):295–302

17. Raychaudhuri S et al (2009) Identifying relationships among genomic disease regions: predicting genes at pathogenic SNP associations and rare deletions. PLoS Genet 5(6):e100053

18. Greco L et al (1998) Lack of correlation between genotype and phenotype in celiac disease. J Pediatr Gastroenterol Nutr 26(3):286–290

19. Gudjonsdottir AH et al (2009) Association between genotypes and phenotypes in coeliac disease. J Pediatr Gastroenterol Nutr 49(2):165–169

20. Fernandez-Cavada-Pollo MJ et al (2013) Celiac disease and HLA-DQ genotype: diagnosis of different genetic risk profiles related to the age in Badajoz, southwestern Spain. Rev Esp Enferm Dig 105(8):469–476

Part II

Laboratory Protocols

Chapter 5

Twenty-Four Hour Ex Vivo Culture of Celiac Duodenal Biopsies

Sarah E.J. Cooper, Sharon Wilson, and Conleth F. Feighery

Abstract

Organ culture is a valuable technique in celiac disease research. It provides the opportunity to examine interactions between different cell types during the disease process without the need for invasive in vivo studies. Biopsies are maintained in an oxygen-rich environment, in contact with, but not submerged in, culture medium. A very straightforward and successful method of organ culture is described here.

Key words Celiac disease, Celiac disease, Biopsy, Culture, Duodenal, Small bowel, Ex vivo, Prolamins

1 Introduction

Browning and Trier first successfully demonstrated the technique for culturing biopsies from adult human intestinal mucosa in 1969 [1]. It was modified from the method used by Trowell in the maintenance of rat organs [2] and remains essentially the same today. The basis of the technique is the maintenance of the biopsy on the surface of the medium rather than submerged in it, as would be the case with cell culture, in an oxygen-rich environment [1].

Organ culture is a particularly useful tool in celiac disease research as there is no proper animal model of this disease, although attempts have been made to produce a model in a number of species including mouse, rat, monkey, and dog [3, 4]. Cell lines have also been widely employed in celiac disease research; however they lack the cell-to-cell interactions that biopsy sections bring [5].

It has been shown that many features of the celiac mucosal immune response are reproduced after 24 h of ex vivo gliadin challenge [6]. Organ culture has been used in celiac disease research to elucidate immune mechanisms of its progression as well as the contributions of various cytokines [6–8]. It has demonstrated the involvement of IL-15 in the progression of the disease [9].

Anthony W. Ryan (ed.), *Celiac Disease: Methods and Protocols*, Methods in Molecular Biology, vol. 1326,
DOI 10.1007/978-1-4939-2839-2_5, © Springer Science+Business Media New York 2015

As biopsies contain a wide variety of cell types, the issues open to investigation are numerous [4]. Organ culture provides a valuable alternative to more invasive in vivo studies and allows control over factors affecting the tissue [2, 9]. Additionally, several biopsies may be obtained from one patient and cultured under a number of different conditions. This greatly increases the number of experiments that can be carried out per patient and also means that each biopsy will have its own, perfectly matched control.

Organ culture can be a difficult technique to perfect. It is important to treat biopsies gently and avoid causing them any undue stress; they will already be suffering some stress as a result of being forcibly removed from their natural surroundings and kept alive in an artificial environment. This and other factors, e.g., the effects of spending 24 h balanced on a piece of mesh, will affect all biopsies that are cultured, regardless of the presence of prolamins etc. Therefore, it is very important, in addition to baseline or "Time 0" controls, to have controls for the 24 h time point that are cultured without the addition of factors such as prolamins.

Following culture, supernatants can be frozen at −20 °C, or −80 °C for more long-term storage, for future analysis, e.g., by ELISA. Biopsies can be processed in a number of ways depending on the analysis required. Possibilities include freezing or formalin fixing and paraffin embedding for future histological analysis or the extraction of RNA for analysis of expression of various elements [6, 8–11].

The technique, as described below, has been used very successfully in this department to investigate the effects of various prolamins on the celiac mucosa [12, 13].

2   Materials

1. Laminar flow hood.
2. Fume hood.
3. Hypoxia chamber (*see* Fig. 1).
4. 95 % O_2/5 % CO_2.

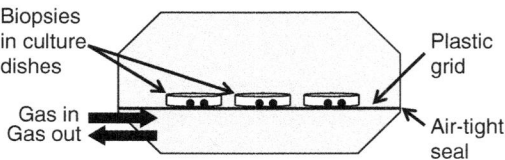

Fig. 1 Hypoxia Chamber: Biopsies in culture dishes are placed in a hypoxia chamber. The chamber is sealed, gassed with 95 % O_2/5 % CO_2 through tubing in the chamber base, and the tubing sealed

5. 37 °C incubator.

6. 60 ml sterile tubs.

7. RPMI⁺⁺: using RPMI containing HEPES and L-glutamine add 15 % heat inactivated, filtered fetal calf serum and 1 % antibiotic/antimycotic.

8. Prolamins: resuspend lyophilised peptic-tryptic digests of gliadin (PT gliadin), avenin, hordein, or secalin in RPMI⁺⁺ under sterile conditions. Allow 2 ml per culture pot. Freeze in 2 ml aliquots at –20 °C (*see* **Note 1**).

9. Hypodermic needles.

10. Sterile organ culture dishes with 2 ml central well.

11. Sterile circular nylon mesh filters of appropriate size to fit central well of organ culture dishes.

12. Sterile Pasteur pipettes.

3   Methods

1. Half fill 60 ml sterile tubs with RPMI (*see* **Note 2**).

2. Place biopsies taken during oesophagogastroduodenoscopy (OGD) directly into tubs of RPMI.

3. In a laminar flow hood, label organ culture dishes and remove lid from first culture dish. Place a mesh filter inside lid of culture dish (*see* **Note 3**). Using a Pasteur pipette carefully remove one biopsy from RPMI taking as little RPMI as possible with the biopsy. Place biopsy on mesh filter and aspirate as much RPMI as possible. Use a hypodermic needle to manipulate biopsy so that it is orientated with the villous side up (*see* **Note 4**).

4. Add 1.5 ml of appropriate culture media to central well of culture dish (*see* **Note 5**).

5. Use a hypodermic needle to carefully transfer mesh to central well of culture dish so that it is floating directly on the media with no bubbles below it (*see* Fig. 2a) (*see* **Note 6**).

6. Use a Pasteur pipette to dot the remainder of the aliquot of culture media around the central well (*see* Fig. 2b) (*see* **Note 7**).

7. Place lid on culture dish and transfer to hypoxia chamber taking care not to disturb delicate balance of biopsy and mesh on media.

8. Repeat **steps 3–7** for all remaining biopsies.

9. Close and seal hypoxia chamber. Carefully transfer chamber to a fume hood, connect to 95 % O_2/5 % CO_2 supply and gas for 5 min (*see* **Note 8**). Turn off gas supply and quickly close clips on tubing.

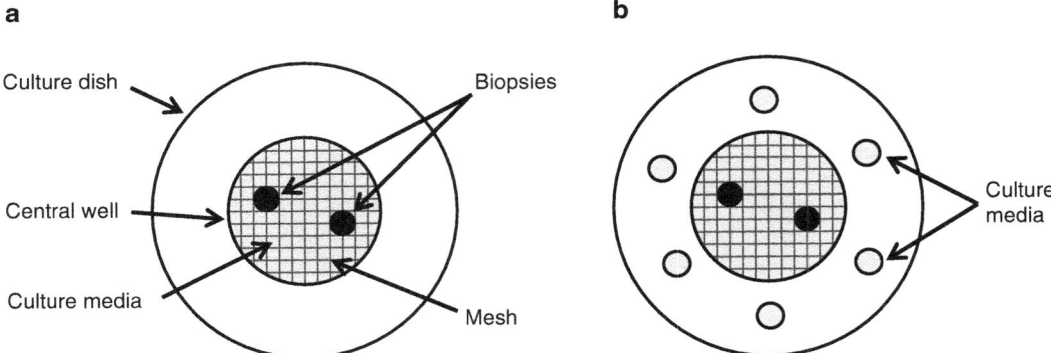

Fig. 2 Biopsies in culture: Biopsies are placed villous side up on a mesh filter. The mesh is then placed in the central well of a culture dish so that it is floating on the surface of the media (**a**). The remaining culture media is dotted around the central well (**b**)

10. Carefully transfer chamber to a 37 °C incubator and leave for 24 h (*see* **Note 9**).

11. After 24 h supernatants can be frozen at –20 or –80 °C. Biopsies can be processed appropriately for the analysis required.

4   Notes

1. We use 5 mg PT gliadin etc. per ml of RPMI^{++}; however other concentrations may be used. It is useful to freeze 2 ml aliquots of RPMI^{++} at this stage for use with control biopsies.

2. Plain RPMI is fine for biopsy collection. More than one biopsy can go into each tub; generally if a pair of biopsies will be cultured together, they are collected in the same tub. It is important not to shock the biopsies or cause them an undue stress; therefore they should not be placed into cold RPMI directly from the fridge. Tubs of RPMI can be brought roughly up to body temperature under a warm tap or alternatively through body heat by being placed in lab coat pocket until use.

3. Mesh filters are often packaged with an identically sized piece of paper between each one; therefore take care not to use a piece of paper in the place of a filter.

4. Biopsy should be manipulated as little as possible and care taken not to damage it with the needle. It is important to get biopsies into culture as quickly as possible. If more than one biopsy is being cultured under the same conditions, they can be cultured in the same dish.

5. Usually at least one biopsy is cultured with just RPMI^{++} to act as a control.

6. It is important that there are no bubbles in the media so that the mesh will be in complete contact with it. If bubbles are present, one hypodermic needle can be used to raise the edge of the mesh and another to remove bubbles.

7. Dotting the remainder of the culture media aliquot around the central well helps to keep the atmosphere humid and prevent the biopsy from drying out during incubation.

8. Make sure clips on both tubes are open before turning on 95 % O_2/5 % CO_2.

9. Incubation can be of shorter duration if desired; however it is difficult to keep biopsies viable for longer than 24 h.

References

1. Browning TH, Trier JS (1969) Organ culture of mucosal biopsies of human small intestine. J Clin Invest 48:1423–1432

2. Howdle PD (1984) Organ culture of gastrointestinal mucosa—a technique for the study of the human gastrointestinal tract. Postgrad Med J 60:645–652

3. Marietta EV, Schuppan D, Murray JA (2009) In vitro and in vivo models of celiac disease. Expert Opin Drug Discov 4:1113–1123

4. Lindfors K, Rauhavirta T, Stenman S, Mäki M, Kaukinen K (2012) In vitro models for gluten toxicity: relevance for celiac disease pathogenesis and development of novel treatment options. Exp Biol Med (Maywood) 237:119–125

5. Stoven S, Murray JA, Marietta EV (2013) Latest in vitro and in vivo models of celiac disease. Expert Opin Drug Discov 8:445–457

6. Maiuri L et al (1996) Definition of the initial immunologic modifications upon in vitro gliadin challenge in the small intestine of celiac patients. Gastroenterology 110:1368–1378

7. Maiuri L et al (2003) Association between innate response to gliadin and activation of pathogenic T cells in coeliac disease. Lancet 362:30–37

8. Fina D et al (2008) Interleukin 21 contributes to the mucosal T helper cell type 1 response in coeliac disease. Gut 57:887–892

9. Maiuri L et al (2000) Interleukin 15 mediates epithelial changes in celiac disease. Gastroenterology 119:996–1006

10. Fais S et al (1992) Gliadin induced changes in the expression of MHC-class II antigens by human small intestinal epithelium. Organ culture studies with coeliac disease mucosa. Gut 33:472–475

11. Mazzarella G et al (2003) An immunodominant DQ8 restricted gliadin peptide activates small intestinal immune response in in vitro cultured mucosa from HLA-DQ8 positive but not HLA-DQ8 negative coeliac patients. Gut 52:57–62

12. Kilmartin C, Lynch S, Abuzakouk M, Wieser H, Feighery C (2003) Avenin fails to induce a Th1 response in coeliac tissue following in vitro culture. Gut 52:47–52

13. Bracken SC, Kilmartin C, Wieser H, Jackson J, Feighery C (2006) Barley and rye prolamins induce an mRNA interferon-gamma response in coeliac mucosa. Aliment Pharmacol Ther 23:1307–1314

Chapter 6

Isolation and Cloning of Gluten-Specific T Cells in Celiac Disease

Yvonne Kooy-Winkelaar and Frits Koning

Abstract

In patients with celiac disease, but not in healthy controls, gluten-specific CD4 T cells are present in the small intestinal lamina propria. Specific stimulation of these T cells due to gluten consumption leads to the release of pro-inflammatory cytokines, in particular IFNγ and IL-21. This leads to tissue damage, the typical morphological alterations like the flattening of the intestinal epithelium, and a variety of disease-associated symptoms including malnutrition, diarrhea, stomach ache, and failure to thrive. Removal of gluten from the diet eliminates the trigger for these CD4 T cells and leads to recovery. These CD4 T cells thus play a crucial role in the disease pathogenesis. Here we describe how such T cells can be isolated and characterized.

Key words Celiac disease, CD4 T cells, Gluten, HLA-DQ2, HLA-DQ8

1 Introduction

Celiac disease (CD) is a small intestinal disorder that with few exceptions only develops in HLA-DQ2 and/or HLA-DQ8 positive individuals [1]. It is now well established that these HLA molecules are uniquely suited to bind and present modified gluten peptides to pro-inflammatory T cells present in the lamina propria of patients with CD. The modification involves the conversion of particular glutamine residues in gluten peptides into the negatively charged glutamic acid by the enzyme tissue transglutaminase (TG2), which induces one or more negative charges in gluten peptides that facilitate high-affinity binding to either HLA-DQ2 or HLA-DQ8 [2, 3]. Typically, HLA-DQ2 binds gluten peptides with a negative charge at position p4 or p6, while HLA-DQ8 binds peptides with a negative charge at p1 and/or p9. Examples of immunodominant gluten peptides are given in Table 1 and in Sollid et al. [4]. However, T cell responses towards native gluten peptides have also been described, particularly in children [5], which may indicate that the T cell response is initiated against

Anthony W. Ryan (ed.), *Celiac Disease: Methods and Protocols*, Methods in Molecular Biology, vol. 1326,
DOI 10.1007/978-1-4939-2839-2_6, © Springer Science+Business Media New York 2015

Table 1
Overview of immunogenic gliadin peptides

DQ2.5-restricted epitopes	
DQ2.5-glia-α1a	P F P Q P E L P Y
DQ2.5-glia-α1b	P Y P Q P E L P Y
DQ2.5-glia-α2	P Q P E L P Y P Q
DQ2.5-glia-α3	FR P E Q P Y P Q
DQ2.5-glia-γ1	P Q Q S F P E Q Q
DQ2.5-glia-γ2	IQ P E Q P A Q L
DQ2.5-glia-γ3	Q Q P E Q P Y P Q
DQ2.5-glia-γ4a	S Q P E Q E F P Q
DQ2.5-glia-γ4b	P Q P E Q E F P Q
DQ2.5-glia-γ4c	Q Q P E Q P F P Q
DQ2.5-glia-γ4d	P Q P E Q P F C Q
DQ2.5-glia-γ5	Q Q P F P E Q P Q
DQ2.5-glia-ω1	I P F P Q P E Q P F
DQ2.5-glia-ω2	P Q P E Q P F P W
DQ8-restricted epitopes	
DQ8-glia-α1	E G S F Q P S Q E
DQ8-glia-γ1a	E Q P Q Q P F P Q
DQ8-glia-γ1b	E Q P Q Q P Y P E

The nine-amino acids core is shown. For a complete listing *see* ref. 4
E: Glutamic acid residues formed by TG2-mediated deamidation of Q residues

native gluten peptides and that this T cell response is amplified by subsequent gluten peptide modification due to the release of TG2 as the result of local tissue damage in the intestine. Thus, when considering the isolation and cloning of T cells from patients with CD (*see* **Note 1**), one should check for T cell responses against both native gluten and TG2 modified gluten.

Gluten is a highly complex mixture of gliadin and glutenin molecules. The gliadins are composed of α-, γ-, and ω-gliadins and the glutenins are divided into low molecular weight (LMW) and high molecular weight (HMW) glutenins, all complex protein families. As the result, gluten isolated from commonly used wheat varieties contains approximately 100 related but distinct proteins. Similarly, barley and rye contain complex mixtures of gliadin-like molecules termed hordeins and secalins respectively. Typically the gliadins are water insoluble. Also, during passage through the gastrointestinal tract, gluten proteins will be partially hydrolyzed by enzymes, pepsin, and trypsin in particular. Specific T cell reactivity towards gluten is therefore determined by making use of pepsin/trypsin treated gluten. Depending on the experiment, this gluten preparation can further be modified by TG2.

2 Materials

2.1 Pepsin-Trypsin Digest of Gluten

Pepsin (Sigma), Trypsin (Sigma), Acetic acid, NaOH, 10 kDa, PBS.

2.2 Tissue Transglutaminase Treatment of Gluten

TG2 (Sigma), $CaCl_2$, triethylamine-acetate.

2.3 Isolation of T Cell Lines and Cloning of T Cell Lines

Culture medium: IMDM: Iscove's Modified Dulbecco's Medium (Bio Whittaker) supplemented with L-glutamine (Gibco) and 10 % (pooled) human serum (*see* **Note 2**).

Cytokines: IL-15 (R&D), IL-2 (Proleukin Novartis).

Phytohaemagglutinin/PHA (HA-).

2.4 Feeders

Irradiated (3000 Rad) peripheral blood mononuclear cells (PBMC) from two unrelated blood donors, isolated by the standard Ficoll-based separation technique and mixed 1:1. One can use either freshly isolated or vitally frozen cells stored in liquid nitrogen. In the latter case it is best to obtain buffy coats, isolate the PBMC, and freeze them down in aliquots that can be used when needed. A buffy coat should yield 300 million cells at least; if frozen down in aliquots of 10 million cells, this will be sufficient for several experiments.

3 Methods

3.1 Preparation of a Pepsin-Trypsin Digest of Gluten

1. Dissolve 1 g of gluten in 10 ml 1 M acetic acid and boil for 10 min.

2. Cool to room temperature and add 10 mg pepsin and incubate for 4 h at 37 °C. Subsequently adjust the pH to 7.8 with 1 M NaOH, add 10 mg trypsin and incubate for 4 h at 25 °C.

3. Finally add 10 mg trypsin inhibitor and dialyze the mixture against 1 l of water.

4. Determine the protein concentration using the BCA assay (Pierce).

5. Store in aliquots at −80 °C.

3.2 Tissue Transglutaminase (TG2) Treatment of Gluten

1. Incubate the pepsin/trypsin digest of gluten (500 µg/ml) or synthetic gluten peptides (500 µg/ml) in 50 mM triethylamine-acetate pH 6.5, 2 mM $CaCl_2$, with TG2 (100 µg/ml; Sigma) at 37 °C for 4 h.

2. Store at –80 °C.

3.3 Generation of T Cell Lines from Small Intestinal Biopsies

1. Use fresh biopsy material, collect this in HBSS medium without Ca^{2+} and Mg^{2+} during endoscopy.

2. Add 1 mM DTT for 10 min followed by 0.75 mM EDTA for 1 h; repeat the EDTA step once. Perform both steps at 37 °C. Transfer the biopsy to 1 ml fresh culture medium (=10 % NHS) in one well of a 24-well plate. Add 80 µg PT-gluten and/or TG2 treated PT-gluten and place in an incubator at 37 °C for 5 days. On day 5 add 20 Cetus units IL-2/ml and 10 ng IL-15/ml to the culture and continue culturing for 5–10 days.

3. Monitor for growth at least every second day. Growth is indicated by the appearance of clusters of "pear-shaped" cells in the cultures. Split the cultures when culture medium changes color from red to yellow or when the bottom is covered with cells.

4. When the cultures change into monolayers of rounded cells, you can either freeze the cells or restimulate the cells by mixing approximately 10^5 biopsy derived T cells with 10^6 irradiated feeder cells in 1 ml culture medium supplemented with 20 Cetus units IL-2/ml, 10 ng IL-15/ml, and 1 µg PHA/ml (Feedermix, see also below). Use as many wells as required to plate all biopsy-derived T cells.

5. Monitor for growth at least every 2 days and split when required.

6. Freeze down cells when the cultures start to become monolayers of rounded cells. Transfer to –80 °C followed by long-term storage in liquid nitrogen.

3.4 Cloning of T Cells from Gluten-Specific T Cell Lines by Limiting Dilution

1. First prepare the feedermix, for twenty 96-well plates you need:
   A total volume of 210 ml containing:

   10 % NHS (=21 ml)

   30 U IL-2/ml (=63 µl from the stock = 100,000 U/ml)

   15 ng IL-15/ml (=315 µl from the stock = 10 µg/ml)

   1.5 µg PHA/ml (=315 µl from the stock = 1 mg/ml)

   1×10^6 irradiated feeder cells/ml (=210×10^6)

   Pipet 100 µl from this mix in the wells of twenty 96-well round-bottom plates.

2. Next make the required dilutions of the T cell line as follows: Count the cells and adjust cell numbers so you reach the following concentration:

0.5×10^6 cells in 2.5 ml culture medium

⇓

100 µl + 9900 µl culture medium (=100 c/50 µl)

⇓

1000 µl + 9000 µl culture medium (=10 c/50 µl)

⇓

3 ml + 100 ml culture medium (=0.3 c/50 µl)

Add 50 µl T cells to the feedermix in the plates in the following scheme:

1 row: 100 cells per well

2 rows: 1 cell per well

Remainder: 0.3 cells/well

3. At day 5 add 50 µl IMDM with 10 % NHS containing 20 Cetus U IL-2/ml and 10 ng IL-15/ml.

4. Score the wells for growth from day 10 onwards.

5. Transfer growing wells to 24-well plates in 1 ml culture medium supplemented with feeders, 20 Cetus units IL-2/ml, 10 ng IL-15/ml, and 1 µg PHA/ml (feedermix).

6. Restimulate expanding clones with feedermix when necessary as described under Subheading 3.3.

7. Freeze down T cell clones when the cultures become monolayers of rounded cells.

8. Transfer to –80 °C followed by long-term storage in liquid nitrogen.

3.5 Alternative Procedures

Cloning by limiting dilution is a widely used method and has the advantage that the full repertoire of T cells specific for a particular antigen is analyzed. However, it is also a very time-consuming method as many of the T cells in a polyclonal line will not be specific for the antigen of interest (*see* **Note 3**). Only a proportion of the T cell clones generated will thus be of interest. Alternatively one can make use of HLA-DQ-gluten tetramers to directly stain and clone gluten-specific T cells [4]. The advantage is that this allows the quantification and isolation of T cells specific for a particular gluten peptide. The disadvantage is that T cells specific for other gluten peptides are not analyzed. In addition, HLA-class II tetramers are difficult to make and not abundant.

Another alternative is to make use of markers detecting T cell activation after stimulation of gluten-specific T cell lines like CD69 or CD25. In combination with FACS sorting, this enriches for T cells specific for the antigen of interest. However, not only gluten-specific T cells will upregulate the expression of such markers as many T cells respond to the cytokines produced due to the antigen-specific stimulation.

4   Notes

1. Although the procedures described above seem straightforward, there are several pitfalls. First, no two T cells are alike, and culturing T cells is a time-consuming process that can only be learned through practice and preferably in a laboratory where there is ample experience with cell culture. Second, everything depends on the quality of the reagents, in particular that of the feeders and the human serum.

2. *Quality of the reagents.* Whatever the source of your human serum, make sure it is of sufficient quality. Some batches are better than others and it is worth testing this. Make sure the complement in the serum is inactivated and the feeders are irradiated (3000 Rad). Make sure the cells are not too old at the start of the procedure, process them immediately, and if they are frozen down, make sure the procedure has gone well by checking cell viability upon thawing of one of the ampoules.

3. *No two T cells are alike.* This means that the above directions cannot be used as a standard protocol where every T cell line and T cell clone is treated identically. Rather it is a guideline that should be adapted depending on the behavior of the line or clone in question. For example, some T cells grow very fast while others grow (much) slower. In practice this means that the former must be split more often and you will likely be able to freeze down many ampoules for future experiments. But if a line/clone grows poorly, a second round of re-stimulation with feedermix may be required before enough cells are obtained to freeze down a few ampoules. It is essential to examine cultures on an almost daily basis to decide which ones are doing well and which ones need attention.

   It is inadvisable to attempt to start this up without access to expert assistance. Seek guidance, preferably within your own institute, from an individual who has experience with culturing of human T cells. If this is not possible, consider spending a few months in a laboratory where there is proper experience.

References

1. Tjon JM, van Bergen J, Koning F (2010) Celiac disease: how complicated can it get? Immunogenetics 62:641–651
2. van de Wal Y, Kooy Y, van Veelen P, Pena S, Mearin L, Papadopulos G, Koning F (1998) Cutting edge: selective deamidation by tissue transglutaminase strongly enhances gliadin-specific T cell reactivity. J Immunol 161: 1585–1588
3. Molberg Ø, McAdam S, Körner R, Quarsten H, Kristiansen C, Madsen L, Fugger L, Scott H, Norén O, Roepstorff P, Lundin KEA, Sjöström H, Sollid LM (1998) Tissue transglutaminase selectively modifies gliadin peptides that are recognized by gut-derived T cells in celiac disease. Nat Med 4:713–717
4. Sollid LM, Qiao SW, Anderson RP, Gianfrani C, Koning F (2012) Nomenclature and listing of celiac disease relevant gluten T-cell epitopes restricted by HLA-DQ molecules. Immunogenetics 64:455–460
5. Vader W, Kooy Y, van Veelen P, de Ru A, Harris D, Benckhuijsen W, Pena S, Mearin L, Drijfhout JW, Koning F (2002) The gluten response in children with recent onset celiac disease. A highly diverse response towards multiple gliadin and glutenin derived peptides. Gastroenterology 122:1729–1737

Chapter 7

Flow Cytometric Analysis of Human Small Intestinal Lymphoid Cells

Margaret R. Dunne

Abstract

Flow cytometry is a powerful technique allowing simultaneous analysis of numerous morphological and phenotypic characteristics of cells and cellular constituents. Improvements in cell isolation techniques in recent years have enabled flow cytometric analyses of cells derived from tissue biopsies. Here we describe a method for isolating and analyzing small intestinal lymphoid cells using flow cytometry.

Key words Small intestine, Intestinal biopsy, Tissue dissociation, Antibodies, Flow cytometry, Intraepithelial lymphocytes, Epithelium, Lamina propria

1 Introduction

Flow cytometry offers numerous advantages over traditional tissue staining approaches, notably in the large and growing number of parameters that can be analyzed simultaneously on diverse cell populations. Markers expressed on the cell surface or within cellular compartments can be readily detected using antibodies conjugated to various fluorescent dyes. This allows for simultaneous evaluation of such diverse parameters as cell phenotype, viability, proliferation, cytotoxicity, growth phase, cytokine production, and even RNA expression, meaning that many experiments can be simultaneously carried out from a limited number of cells. As endoscopic procedures have improved, fewer and smaller small intestinal biopsies are taken, which can limit the amount of tissue available for research purposes. Flow cytometric analysis typically does not require a very large number of cells [1] and can potentially generate a lot of data quite rapidly. This makes for an attractive and robust approach for immunological analysis of tissue cells, useful for studying rare cell populations, diagnostic biomarkers and disease monitoring, as well as presenting a powerful complement to histological methods. One limitation to flow cytometric analysis

Anthony W. Ryan (ed.), *Celiac Disease: Methods and Protocols*, Methods in Molecular Biology, vol. 1326, DOI 10.1007/978-1-4939-2839-2_7, © Springer Science+Business Media New York 2015

lies in the requirement for a single-cell suspension. As a result, various approaches have been devised in order to isolate cells from tissue, combining mechanical and enzymatic means, while minimizing deleterious effects on cells [2–5]. Here we present an optimized protocol for the isolation and flow cytometric analysis of viable cells from human small intestinal tissue biopsies and discuss challenges and common pitfalls.

2 Materials

2.1 Cell Dissociation Solutions

1. Calcium- and magnesium-free Hank's Balanced Salt Solution (HBSS) supplemented with 5% v/v foetal calf serum (FCS).

2. Dissociation buffer: Collection buffer supplemented with 1 mM ethylenediaminetetraacetic acid (EDTA) and 1 mM dithiothreitol (DTT).

3. Complete RPMI (cRPMI): RPMI 1640 with Glutamax, supplemented with 10 % v/v FCS, 100 U/ml penicillin, 100 μg/ml streptomycin, 2.5 μg/ml amphotericin B Fungizone, and 0.02 M HEPES.

4. Collagenase solution: 130 U/ml collagenase (type IV) in cRPMI buffer (see Notes 1 and 2).

5. PBA buffer: 1 % v/v bovine serum albumin, 0.02 % v/v sodium azide in phosphate buffered saline, adjusted to pH 7.

6. 0.5 % Paraformaldehyde (PFA) solution.

3 Methods

It is imperative that tissue biopsies are collected and processed promptly, as viability, particularly of epithelial cells, rapidly declines in vitro [2]. All tissue buffer solutions should be warmed to 37 °C prior to use. Aseptic technique should be adhered to throughout the procedure. Solutions containing enzymes should be made up freshly and used as quickly as possible as, depending on type and source, some of these reagents have quite short activity half-lives. Solutions may be prepared in concentrated aliquots and frozen for future use, but freeze-thaw cycles should be avoided.

3.1 Tissue Collection and Preparation

1. Collect biopsies into a sterile container containing calcium- and magnesium-free HBSS supplemented with 5 % FCS.

2. Wash tissue fragments by carefully removing buffer using a Pasteur pipette and add approximately 25 ml fresh HBSS buffer. Agitate tissue by gently pipetting up and down or inverting the capped container. Allow tissue fragments to settle, pipette off supernatant, and repeat washing in this way three times.

3. If necessary, cut tissue into pieces approximately 5 mm in size, using a sterile scalpel.

3.2 Tissue Processing

1. Remove HBSS buffer and resuspend tissue in dissociation buffer.

2. Place tube containing biopsy material into a shaker, and set to maximum agitation at 37 °C for 1 h (*see* **Note 3**).

3. Strain dissociated tissue through a sterile 40 μm nylon mesh cell strainer into a sterile 50 ml tube. Wash through with cRPMI buffer.

4. Carefully remove remaining solid biopsy tissue fragments from the filter using sterile forceps and place into a tube of collagenase solution. Secure tube to a shaker in a 37 °C incubator for 30 min–3 h (*see* **Note 4**), checking tissue dissociation progress periodically.

5. Centrifuge the eluent, containing the epithelial cell suspension, at $300 \times g$ for 7 min. Discard supernatant and count epithelial cells.

6. Wash and resuspend cells in an appropriate amount of cRPMI, and remove an aliquot for viability counting (*see* **Note 5**).

3.3 Cell Staining

1. Resuspend cells in PBA buffer at a concentration of 2×10^6 cells/ml, and add 100 μl to each FACS tube (*see* **Note 6**).

2. Add flow cytometry antibodies and incubate for 15 min in the dark at room temperature (*see* **Notes 7** and **8**).

3. Wash cells in 2 ml PBA, and centrifuge at $300 \times g$ for 7 min.

4. Discard supernatant, vortex cell pellet gently, and fix cells by incubating for 15 mins in 0.5 % PFA.

5. Wash and resuspend cells in approximately 0.5 ml PBA.

3.4 Cell Acquisition

1. Cells should be acquired as soon as possible after staining (*see* **Note 9**). Typical staining patterns for small intestinal epithelial and lamina propria derived cells are shown compared to whole blood in Fig. 1.

4 Notes

1. Collagenase activity can be inhibited by EDTA, DTT, and other factors but is promoted in the presence of calcium ions. Ensure tissue is washed thoroughly in cRPMI prior to collagenase incubation step.

2. Tissue clumping can occur due to cell damage and DNA release. This may be minimized by addition of DNase I to collagenase solution [6].

3. A constant high level of tissue agitation is required for optimal epithelial cell disaggregation. To further increase agitation, place the tube containing the tissue into a larger container secured to the shaker [7].

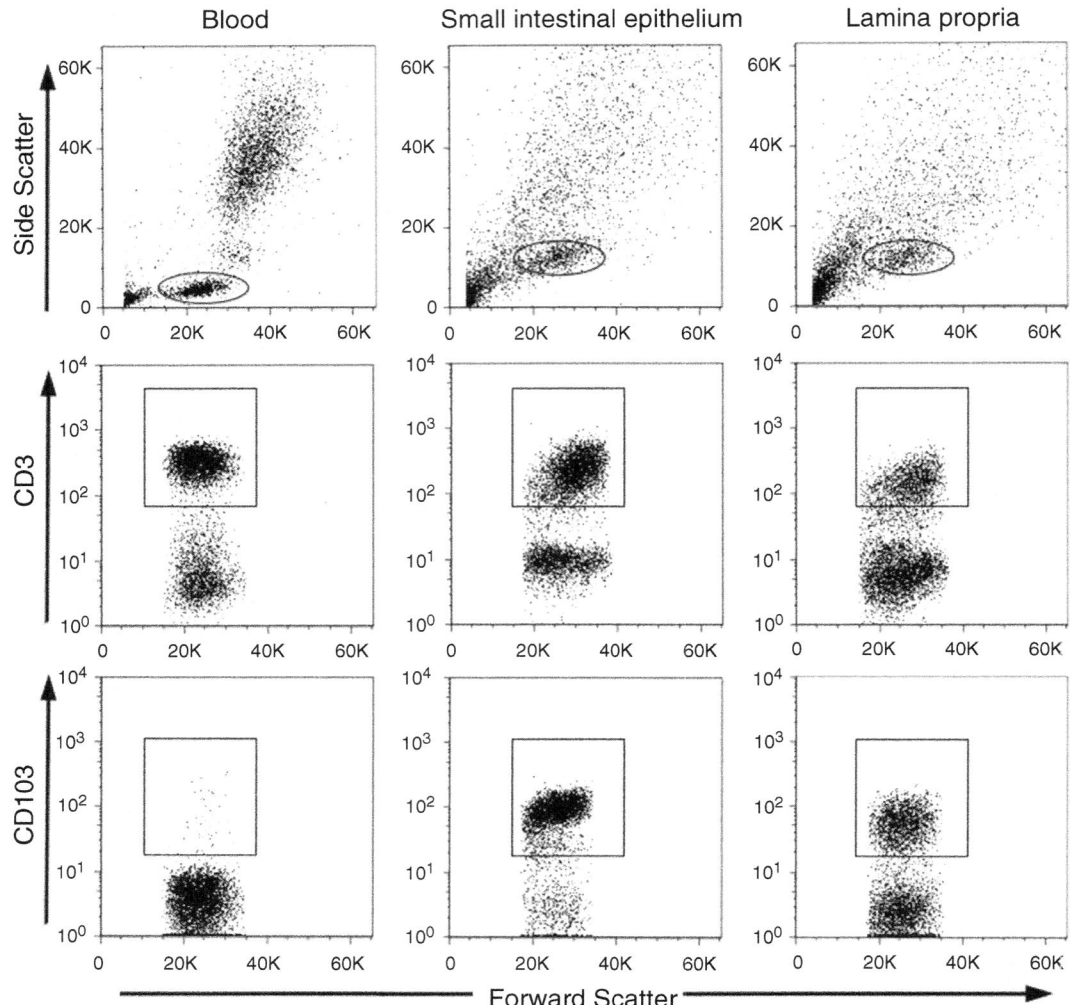

Fig. 1 Flow cytometry dot plots show typical staining patterns for cells isolated from whole blood, small intestinal epithelium, and lamina propria. Forward scatter (FSC) and side scatter (SSC) properties are shown for each compartment in the *first row* of plots, CD3 expression in the *middle row*, and CD103 expression in the *bottom row*

4. Collagenase and other contaminating proteases can cleave cell surface markers. These effects may be minimized by reducing incubation time. Not all cell surface markers are affected in this manner [4, 8], but researchers are advised to carry out time-course experiments to assess the effects of specific collagenase solutions on particular markers of interest. Studies also suggest that prolonged exposure to collagenase affects cell viability [6]; therefore it is recommended to keep collagenase exposure time to a minimum and wash cells thoroughly after incubation.

5. Dead cells may be removed by Ficoll or Percoll gradient centrifugation at this point [2, 7], although this may cause significant cell loss and may only be practical when a lot of cells are available. Otherwise, inclusion of a viability dye with flow cytometry staining is strongly recommended, as dead cells and debris may significantly skew data.

6. Alternatively, at this point cells may be cultured or stimulated in vitro and intracellular components (transcription factors, cytokines, etc.) can be assessed using intracellular staining protocols [9].

7. As well as intestinal cells stained for markers of interest, it is important to include additional control tubes, e.g., unstained intestinal cells are useful for identification of autofluorescence, commonly encountered in tissue-derived material. Isotype and "fluorescence minus one" (FMO) controls are also strongly recommended in order to define correct staining patterns, as these may vary significantly to those usually seen for blood cells.

8. Incorporation of the lineage marker CD45 is recommended for accurate identification and gating of cells of hematopoietic origin. Intraepithelial lymphocyte populations from different intestinal compartments can be assessed for purity by staining for the epithelial-homing integrin CD103, or CD7 expressed by bone marrow-derived cells [7, 10, 11]. This method is superior to CD3 staining, as a significant proportion of intraepithelial gut lymphocytes are typically CD3 negative [7].

9. Counting beads can also be added to stained cells prior to acquisition, allowing enumeration of cell populations.

References

1. Leon F (2011) Flow cytometry of intestinal intraepithelial lymphocytes in celiac disease. J Immunol Methods 363:177–186. doi:10.1016/j.jim.2010.09.002

2. Bull DM, Bookman MA (1977) Isolation and functional characterization of human intestinal mucosal lymphoid cells. J Clin Invest 59:966–974. doi:10.1172/JCI108719

3. Madrigal L, Lynch S, Feighery C et al (1993) Flow cytometric analysis of surface major histocompatibility complex class II expression on human epithelial cells prepared from small intestinal biopsies. J Immunol Methods 158: 207–214

4. Abuzakouk M, Feighery C, O'Farrelly C (1996) Collagenase and Dispase enzymes disrupt lymphocyte surface molecules. J Immunol Methods 194:211–216

5. O'Keeffe J, Doherty DG, Kenna T et al (2004) Diverse populations of T cells with NK cell receptors accumulate in the human intestine in health and in colorectal cancer. Eur J Immunol 34:2110–2119. doi:10.1002/eji.200424958

6. Kralovánszky J, Harrington F, Greenwell A, Melnick R (1990) Isolation of viable intestinal epithelial cells and their use for in vitro toxicity studies. In Vivo (Athens, Greece) 4:201–204

7. Leon F, Roy G (2004) CD7 and CD103 as markers for the clinical enumeration of small-bowel intraepithelial lymphocytes. Scand J Gastroenterol 39:802. doi:10.1080/00365520410003290, author reply 803

8. Mulder WMC, Koenen H, Muysenberg AJC et al (1994) Reduced expression of distinct T-cell CD molecules by collagenase/DNase treatment. Cancer Immunol Immunother 38:253–258. doi:10.1007/BF01533516

9. Dunne MR, Elliott L, Hussey S et al (2013) Persistent changes in circulating and intestinal γδ T cell subsets, invariant natural killer T cells

and mucosal-associated invariant T cells in children and adults with coeliac disease. PLoS One 8(10), e76008

10. Cerf-Bensussan N, Jarry A, Brousse N et al (1987) A monoclonal antibody (HML-1) defining a novel membrane molecule present on human intestinal lymphocytes. Eur J Immunol 17:1279–1285. doi:10.1002/eji.1830170910

11. Eiras P, Leon F, Camarero C et al (2000) Intestinal intraepithelial lymphocytes contain a CD3–CD7+ subset expressing natural killer markers and a singular pattern of adhesion molecules. Scand J Immunol 52:1–6

Chapter 8

Adaptation of a Cell-Based High Content Screening System for the In-Depth Analysis of Celiac Biopsy Tissue

Sarah E.J. Cooper, Bashir M. Mohamed, Louise Elliott, Anthony Mitchell Davies, Conleth F. Feighery, Jacinta Kelly, and Jean Dunne

Abstract

The IN Cell Analyzer 1000 possesses several distinguishing features that make it a valuable tool in research today. This fully automated high content screening (HCS) system introduced quantitative fluorescent microscopy with computerized image analysis for use in cell-based analysis. Previous studies have focused on live cell assays, where it has proven to be a powerful and robust method capable of providing reproducible, quantitative data. Using HCS as a tool to investigate antigen expression in duodenal biopsies, we developed a novel approach to tissue positioning and mapping. We adapted IN Cell Analyzer 1000's image acquisition and analysis software for the investigation of tissue transglutaminase (tTG) and smooth muscle alpha-actin (SM α-actin) staining in paraffin-embedded duodenal tissue sections from celiac patients and healthy controls. These innovations allowed a quantitative analysis of cellular structure and protein expression. The results from routine biopsy material indicated the intensity of protein expression was altered in celiac disease compared to normal biopsy material.

Key words High content screening (HCS), Celiac disease (CD), Tissue transglutaminase (tTG), Smooth muscle alpha-actin (SM α-actin)

1 Introduction

1.1 IN Cell 100 (GE Technology) Background

The cell-based high content screening (HCS) system operates on the principle of fully automated fluorescence microscopy. This technology was introduced as a modern drug discovery tool and represents a major breakthrough in bringing quantitative fluorescence microscopy to bear on the automation of cell biology and computerized image analysis. It provides a fast and convenient means of conducting multiparametric characterization of multiple biological responses through simultaneous assessment of a series of molecular and cellular targets [1–6]. This may include monitoring

Anthony W. Ryan (ed.), *Celiac Disease: Methods and Protocols*, Methods in Molecular Biology, vol. 1326,
DOI 10.1007/978-1-4939-2839-2_8, © Springer Science+Business Media New York 2015

subcellular localization and redistribution of individual proteins within complex cellular structures such as organelles [1–6]. This technology also offers clear advantages to more traditional biochemical or genetic analysis. HCS can monitor and characterize physiological responses within the context of the structural and functional networks of cells in both normal and diseased states [7–9]. It has been used primarily in the context of cell suspensions, and its use in tissue analysis is only emerging [1, 2].

1.2   Advantages Over Traditional Methods

The IN Cell 1000's greatest advantage is its automated fluorescence microscope based image acquisition system that allows the capture of several images at a much higher resolution and at a higher rate compared to that of a normal immunofluorescence microscope. The data analysis software produces a reproducible, quantitative measurement of staining intensity eliminating observer bias and subjectivity.

1.3   Adaption for Use with Tissue

We introduced positioning and mapping modifications that facilitated the location of tissue by the IN Cell 1000 Analyzer (Fig. 1). The preview scan mode was used to quickly locate the region of interest prior to acquisition, enabling the acquisition of only those fields that contained tissue, thereby increasing acquisition speed (Fig. 2a–d). Details of cover slip thickness, location of section, sample area size to be acquired, and distance between areas were entered into the acquisition software (Fig. 1c).

In a published study [7] duodenal biopsies from celiac patients and healthy controls were compared. The intensity and co-localization of two proteins of interest, SM α-actin and tTG, were measured in stellate, pericryptal myofibroblast. The celiac patients had varying levels of duodenal lesions (Marsh lesions I, II, III, and IV) [10]. IN Cell 1000 facilitated simultaneous acquisition of three fluorescent staining patterns: nuclear staining, SM α-actin, and tTG (Fig. 3a–c). Overlaying the acquired (Fig. 3d) regions yielded a fused image allowing for accurate interpretation of the co-localization of these proteins in the cells under study—the pericryptal myofibroblasts. Finally, all acquired microscopic fields were pieced together to create a jigsaw like image of the tissue section providing an overall picture of the pattern distribution for each target protein.

The impressive analysis software of the IN Cell 1000 workstation allows the development of sensitive and specific protocols that can be easily altered to fit the user's individual requirements. For instance, in a reported study [7] by defining certain measures relating to cell shape, the analyzer specifically and accurately recognized and selected the myofibroblasts for analysis. Furthermore we were

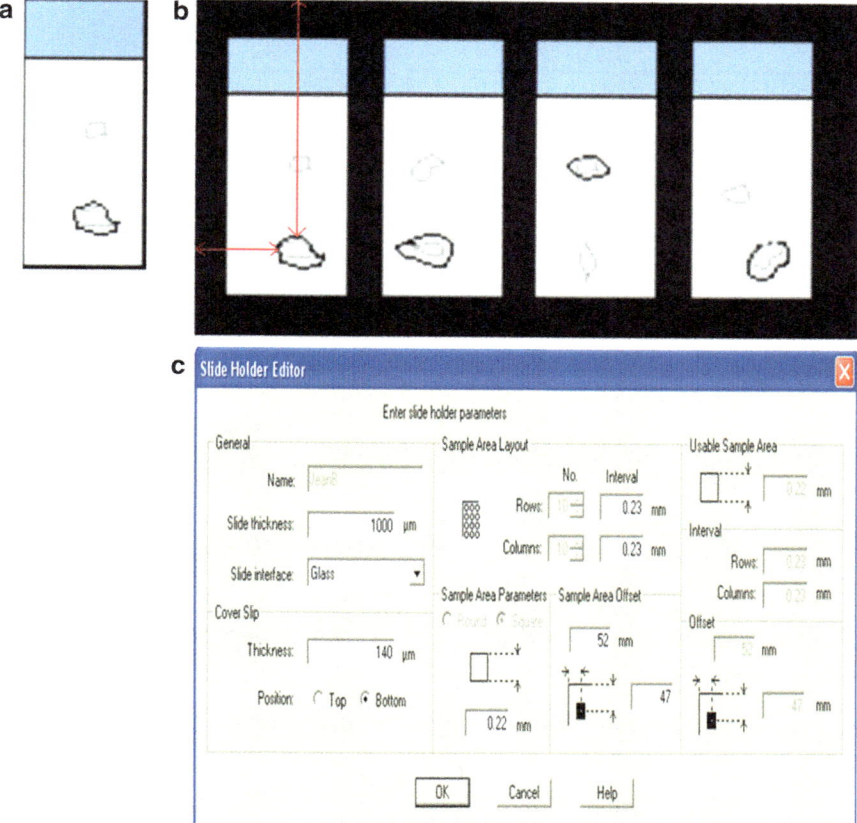

Fig. 1 Tissue Positioning: Tissue sections were circled with black permanent marker (**a**), slides were placed face down in the slide holder and the distance in millimeters from the top left corner of the slide holder to the tissue section measured (*red arrows*) (**b**). Relevant details were entered into the acquisition software for each section (**c**)

able to select multiple parameters such as cell count, cell area, SM α-actin intensity, and tTG intensity that provided us with additional detailed information within each selected cell. The derived numerical data from the processed images allowed us to make an unbiased decision on whether there was a significant difference in the results obtained from normal controls and celiac patients (Fig. 4a–d).

Our novel approach together with modifications to the HCS acquisition protocol using image analysis tool box allowed us to demonstrate a new application for this system in scanning and analysis of fixed tissue sections.

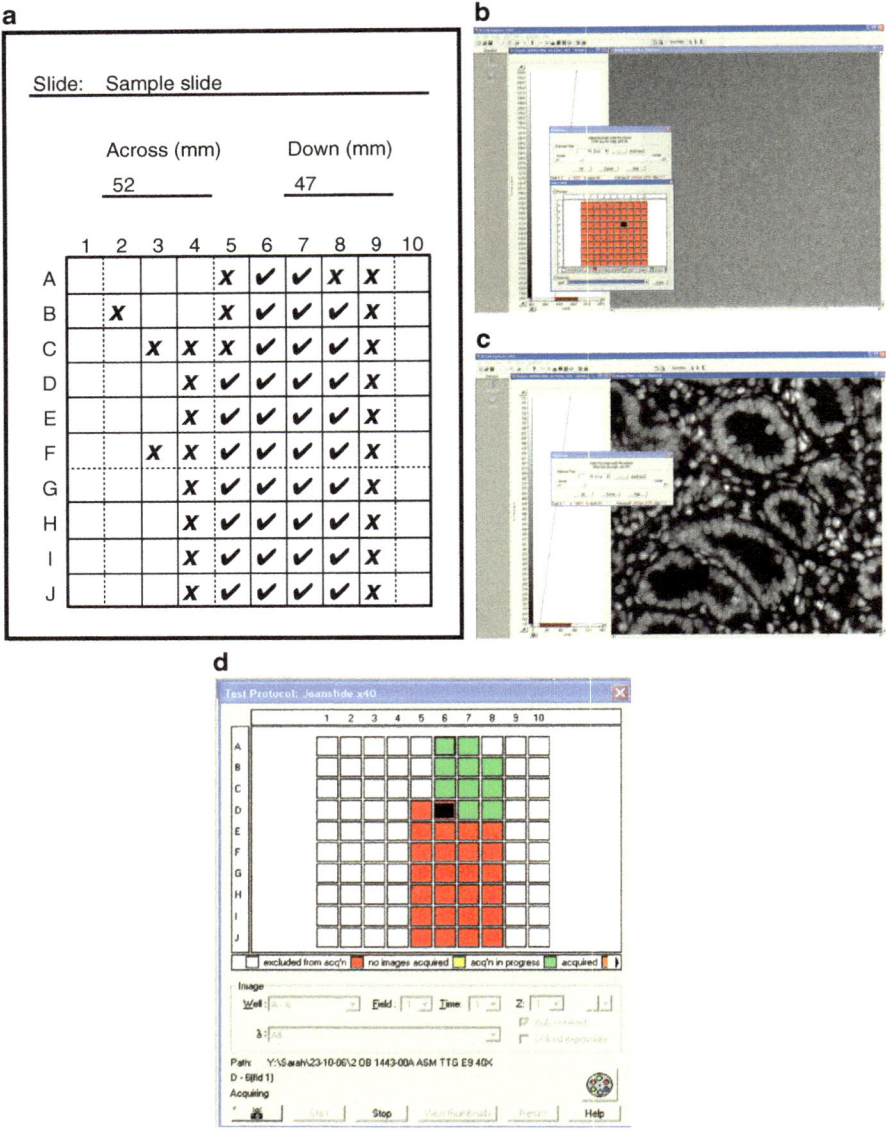

Fig. 2 Tissue Mapping: Excel grid showing slide identity, tissue location, empty sample areas (X), and sample areas containing tissue (✓□) (**a**), empty sample area (**b**), sample area with tissue (**c**), screenshot taken during acquisition showing acquired areas (*green*), areas to be acquired (*red*), area currently being acquired (*black*), and sample areas without tissue that have been eliminated (*blank squares*) (**d**)

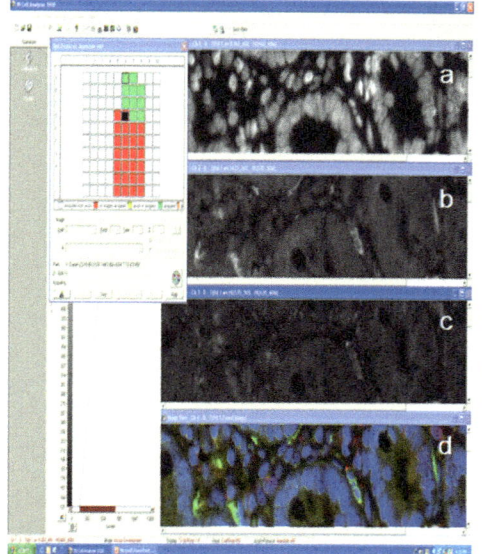
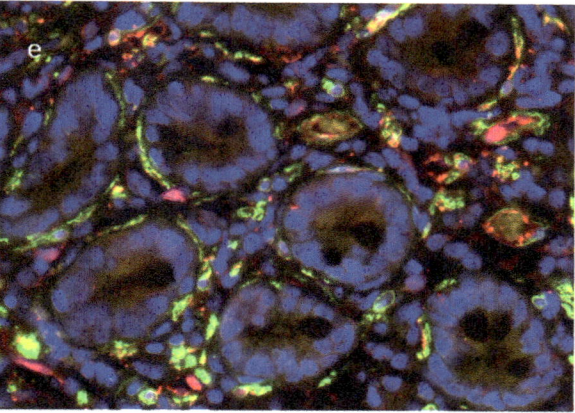

Fig. 3 For the image acquisition protocol of triple color fluorescent tissue images, Channel 1 (D360/40—HQ460/40) was designated as DNA binding dye/Hoechst nuclear staining (**a**), Channel 2 (S475/20—HQ535/50) was designated as "nuclear" or SM α-actin staining (**b**), and channel 3 (HQ535/50—HQ620/60) was designated as "cell" or tissue transglutaminase (tTG) staining (**c**). A colored overlay image of all three channels is shown (**d** and **e**)

2 Materials

2.1 Immuno-fluorescent Staining

1. Paraffin-embedded duodenal tissue, glass slides, microtome, and oven (*see* **Note 1**).

2. Two baths of xylene and a series of graded alcohols: 100, 70, and 50 % (*see* **Note 2**).

3. Phosphate-buffered saline solution (PBS) at pH 7.4.

4. Citrate buffer: dissolve appropriate quantity of citric acid for a 0.01 M final solution in 900 ml dH$_2$O and adjust pH to 6.0 using NaOH. Make up to 1 L with dH$_2$O (*see* **Note 3**).

5. Desired primary antibodies, in this case anti-tissue transglutaminase (tTG) and anti-smooth muscle alpha-actin (SM α-actin), as well as appropriate fluorescent secondary antibodies (*see* **Note 4**).

6. Dilution buffer: 0.01 % sodium azide and 0.15 % BSA in PBS.

7. Humidity chamber.

8. DNA binding dye diluted in dH$_2$O to a concentration of 1 μM.

9. Fluorescent mounting media and cover slips.

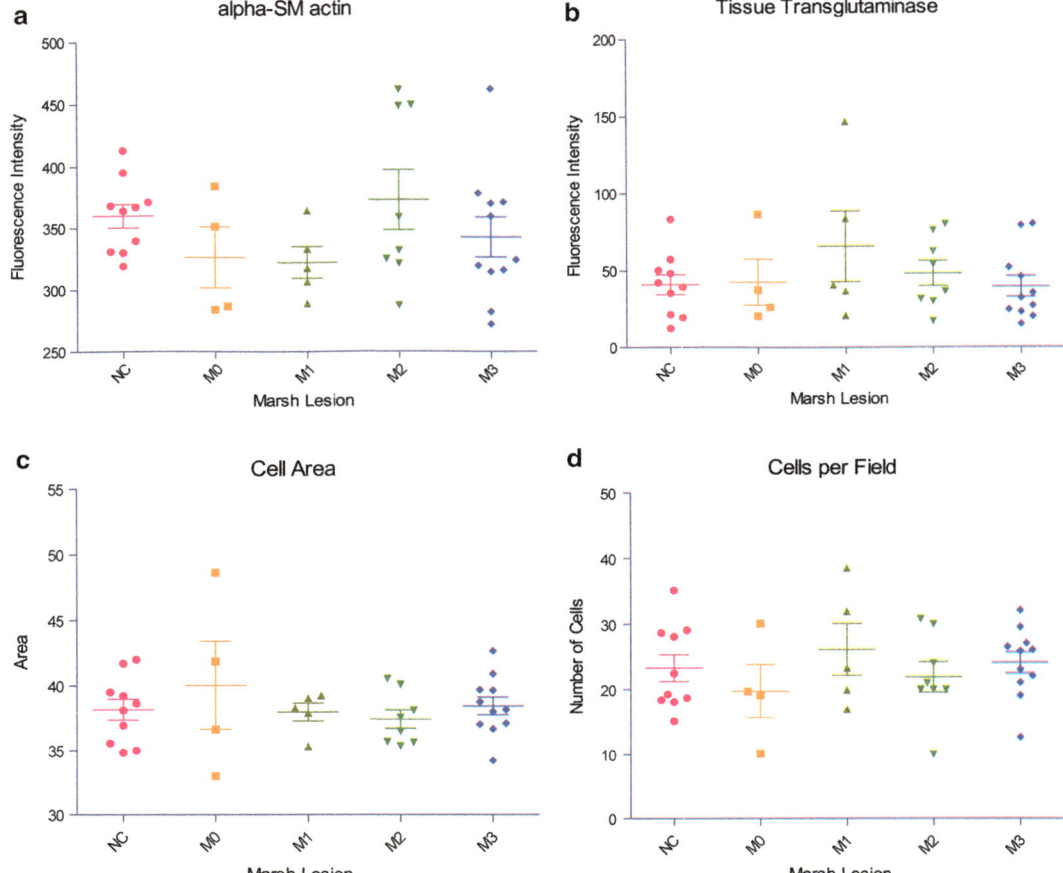

Fig. 4 Fluorescence intensity of smooth muscle α-actin (**a**) and tTG (**b**) in pericryptal myofibroblasts from control (NC) and celiac biopsies with Marsh lesions ranging from M0 (normal villous architecture) through M1 (lymphocyte infiltration) and M2 (villous blunting) to M3 (total villous atrophy). Myofibroblast cell area (**c**) and cells numbers (**d**) expressing smooth muscle α-actin in control and celiac biopsies

2.2 IN Cell Analysis

1. Access to IN Cell Analyzer with slide holder.
2. Black permanent marker and ruler (*see* **Note 5**).
3. Grid representing the potential tissue sample area (*see* Fig. 2a).

3 Methods

3.1 Immuno-fluorescent Staining

1. Cut 2 μm thick duodenal sections from paraffin-embedded blocks using a microtome. Place on glass slides and dry overnight at 60 °C.
2. Place slides in slide rack. Dewax and rehydrate by placing slide rack in the following baths: xylene I and xylene II for 10 min each, 100, 70, and 50 % alcohol, dH$_2$O and PBS, all for 5 min (*see* **Note 6**).

3. Use blank slides to fill any gaps in slide rack. Place in microwaveable plastic tub and add 800–900 ml citrate buffer so that slide rack is well submerged. Wrap container completely in two layers of cling film and pierce cling film several times at open surface of tub. Microwave on highest power setting for 25 min and leave to cool in microwave for 20 min. Transfer tub to sink and leave under running cold tap H_2O for 5 min then move slide rack into a bath of PBS for 5 min (*see* **Note 7**).

4. Prepare all antibodies at appropriate concentrations in dilution buffer. Make 50 µl of antibody dilution per slide.

5. Add 50 µl of first primary antibody, cover with cover slip, and incubate in humidity chamber for 1 h at room temperature (RT) (*see* **Note 8**).

6. Float off cover slip in a bath of PBS, transfer slides to slide rack and wash 2×3 min in PBS.

7. Repeat **steps 5** and **6** for all remaining antibodies.

8. Counterstain with DNA binding dye for 5 min at RT (*see* **Note 9**).

9. Repeat **step 6**.

10. Mount with fluorescent mounting media and cover slip. Leave in a cardboard slide holder overnight at RT to solidify then store @ 4 °C (*see* **Note 10**).

3.2 IN Cell Analysis

In order to use the IN Cell Analyzer 1000 for the analysis of fluorescent staining intensity in tissue sections, we used a novel method of tissue positioning that facilitated image capture, together with a commercially produced slide holder which we had previously developed for use with Cellomics [8]. Developer Toolbox was used to design a protocol for the analysis of triple color fluorescent staining patterns in the pericryptal myofibroblasts of duodenal biopsies from healthy controls and celiac patients [7].

3.2.1 Tissue Positioning

1. Turn slides face down and outline the sections of interest using black permanent marker (*see* Fig. 1a).

2. Place slides face down in slide holder and measure the distance of each section from the top left corner of the plate in millimeters across and down (*see* Fig. 1b) (*see* **Notes 11** and **12**).

3. Transfer slide holder to IN Cell Analyzer (*see* **Note 12**).

4. Enter details such as cover slip thickness, location of section, sample area parameters, and sample area interval into the acquisition software (*see* Fig. 1c) (*see* **Note 13**).

3.2.2 Tissue Mapping

1. Set excitation and emission filters for blue/UV light (D360/40—HQ460/40), green fluorescent light (S475/20—HQ535/50), and red fluorescent light (HQ535/50—HQ620/60) in channels 1, 2, and 3 respectively.

2. Using channel 1 (blue light), check individual squares within the potential acquisition area for presence of tissue.

3. On the sample area grid (*see* Fig. 2a), mark empty squares (*see* Fig. 2b) with an x and squares containing tissue (*see* Fig. 2c) with a tick until the perimeter of the section is mapped (*see* **Note 14**).

4. Using the software, select squares containing tissue for acquisition and eliminate all others (*see* Fig. 2d).

3.3 Acquisition Protocol

Problems with focusing were overcome by altering the cover slip thickness or exposure time. For additional focus the "software auto focus" function was activated so that each square was focused individually during acquisition. This increased acquisition time but produced a clearer image. Excitation and emission filters were specified for blue/UV light (D360/40—HQ460/40), FITC (S475/20—HQ535/50), and TRTIC (HQ535/50—HQ620/60). These included a DAPI filter (channel 1), which detected blue fluorescence indicating nuclear intensity; FITC filter (channel 2), which detected green fluorescence indicating smooth muscle α-actin; and a TRITC filter (channel 3), which detected tTG changes with red fluorescence signals (Fig. 3). The ×40 objective and Trichroic filter were used in image acquisition using the optimized acquisition protocol outlined above.

3.4 Image Analysis: IN Cell Investigator Developer Toolbox

Developer Toolbox is used in high content analysis applications where pre-developed image analysis modules are not suitable. A selection of advanced segmentation, pre-processing, and post-processing tools provides full control over the sequence of steps in analysis routines.

For the present analysis we were measuring three color staining of elongated, stellate myofibroblasts in the pericryptal region of duodenal biopsies. Dual Area Object analysis, allowing the simultaneous measurement of subcellular inclusions, was chosen as the most appropriate analysis method in the IN Cell Investigator Developer Toolbox. *Channels 1, 2, and 3* were designated as per the acquisition protocol (Fig. 3). Areas of the duodenal tissue containing crypts were chosen for analysis and identified in the thumbnail image.

For nuclear (SM α-actin or green channel) analysis "Region Growing" was chosen; for cell (tissue transglutaminase or red channel) analysis "Multiscale Tophat" (minimum area and sensitivity were set).

Measurements designated in the nuclear channel were "Nuclear Area," "Nuclear/Cell Intensity," "Nuclear Elongation," and "Nuclear Intensity." Nuclear area of 20 μm was chosen as the filter setting. Measurements designated in the cell channel were "Area," "Elongation," and "Intensity."

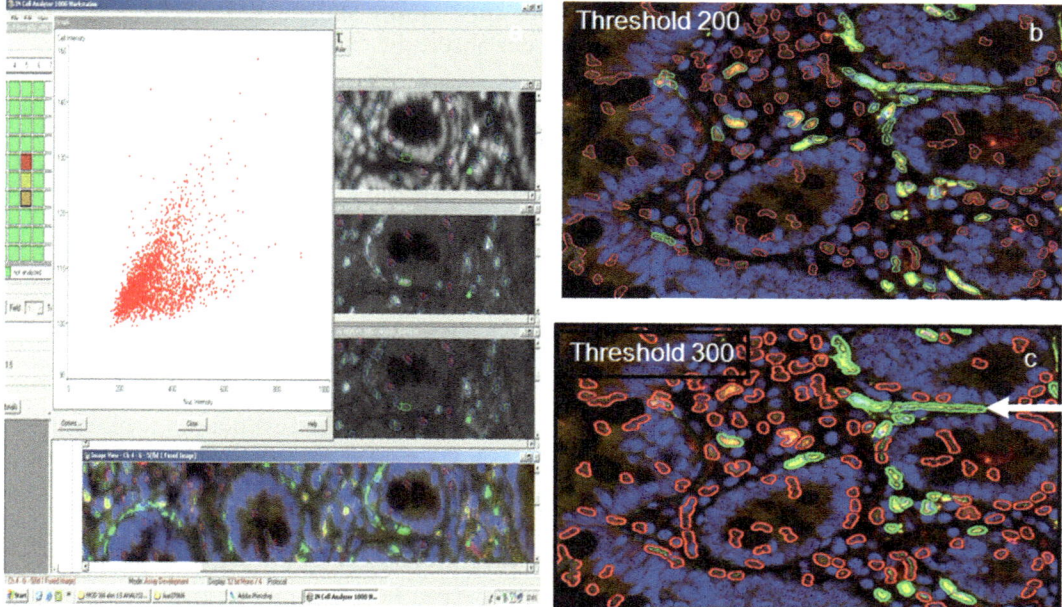

Fig. 5 Threshold Settings: Optimized threshold settings for stellate myofibroblasts staining were estimated using a dot plot of smooth muscle α-actin FITC (nuclear intensity channel 2) versus tTG (cell intensity channel 3) (**a**). Results shown for green channel (channel 2) threshold settings 200 (**b**) and 300 (**c**) indicate the successful inclusion of stellate pericryptal myofibroblasts expressing smooth muscle α- actin (green staining) (*white arrow*)

3.5   Dual Object Area Analysis

The protocol was developed to examine our target cells, which had an elongated morphology and were stained with smooth muscle α-actin. Tissue sections present a variety of cells with individual staining patterns and morphologies, so in targeting one cell type the protocol must include those cells of interest whilst excluding all others. Threshold settings were optimized to exclude cells that did not show an elongated morphology or staining with anti-SM α-actin (Fig. 5).

4   Notes

1. Electrostatic slides are best, as sections will easily float off plain slides during staining process.

2. 70 and 50 % alcohols can be made by adding appropriate quantities of dH_2O.

3. pH will have to be raised significantly; therefore NaOH pellets are useful for initial adjustments.

4. Data sheets for fluorescent antibodies rarely specify suitability for staining of formalin-fixed paraffin-embedded tissue.

However, we have successfully stained such tissue with many different fluorescent antibodies using this protocol.

5. Always use a black marker, colored ink will autofluoresce.

6. Allow liquid to drain from rack and blot on tissue paper after each bath to prevent contamination of the next bath. Xylenes and alcohols can be reused but should be changed as soon as they become visibly dirty, i.e., bits of wax floating around.

7. Adding blank slides to fill any empty slots in the slide rack is important to equalize heat distribution. A high powered microwave is necessary for proper antigen retrieval. With age a microwave may lose power and no longer be suitable for antigen retrieval. It is important that the citrate buffer boils for proper retrieval. Tub will be very hot even after cooling in the microwave. When tub is under running tap H_2O the flow of water should not be directed at the sections as they may separate from the slides.

8. Some antibodies produce better results when incubated at 4 °C overnight in the dark. For negative controls, PBS is used in the place of the primary antibody.

9. DNA binding dye visualizes the nuclei of cells. This is extremely useful when locating a tiny section during analysis. It can also be useful in determining the location of particular antibodies within a cell.

10. Slides must be protected from light to preserve fluorescence of antibodies; therefore slide folders with clear plastic covers are not suitable for use at this stage.

11. Slides are placed into the holder face down so that analysis is through the cover slip rather than the entire thickness of the slide.

12. As there is room for slides to move around within the slide holder, it is important slides are kept tight to one corner while measuring and transferring to IN Cell Analyzer.

13. Location of section relative to top left corner of slide holder has to be entered in the "Sample Area Offset" box before acquisition of each individual section (Fig. 1c). The potential acquisition area consists of a grid of 10×10 squares. "Sample Area Parameters" refers to the size of each individual square (Fig. 2d). A distance of at least 0.01 mm must be left between each individual square. "Interval" refers to the distance from the center of one square to the center of the next square; therefore the minimum value is the size of an individual square plus 0.01 mm.

14. Acquisition of images is time consuming and uses a lot of memory. Therefore, it is necessary to determine which squares of the potential acquisition area actually contain tissue so that only those are acquired (Fig. 2).

References

1. Ghosh RN, DeBiasio R, Hudson CC, Ramer ER, Cowan CL, Oakley RH (2005) Quantitative cell-based high-content screening for vasopressin receptor agonists using transfluor technology. J Biomol Screen 10(5):476–484. doi:10.1177/1087057105274896

2. Ghosh RN, Lapets O, Haskins JR (2007) Characteristics and value of directed algorithms in high content screening. Methods Mol Biol (Clifton, NJ) 356:63–81

3. Mohamed BM, Movia D, Knyazev A, Langevin D, Davies AM, Prina-Mello A, Volkov Y (2013) Citrullination as early-stage indicator of cell response to single-walled carbon nanotubes. Sci Rep 3:1124. doi:10.1038/srep01124

4. Mohamed BM, Verma NK, Davies AM, McGowan A, Crosbie-Staunton K, Prina-Mello A, Volkov Y (2012) Citrullination of proteins: a common post-translational modification pathway induced by different nanoparticles in vitro and in vivo. Nanomedicine (Lond) 7(8):1181–1195. doi:10.2217/nnm.11.177

5. Mohamed BM, Verma NK, Prina-Mello A, Williams Y, Davies AM, Bakos G, Volkov Y (2011) Activation of stress-related signalling pathway in human cells upon SiO2 nanoparticles exposure as an early indicator of cytotoxicity. J Nanobiotechnol 9:29. doi:10.1186/1477-3155-9-29

6. Towne DL, Nicholl EE, Comess KM, Galasinski SC, Hajduk PJ, Abraham VC (2012) Development of a high-content screening assay panel to accelerate mechanism of action studies for oncology research. J Biomol Screen 17(8):1005–1017. doi:10.1177/1087057112450050

7. Cooper SEJ, Kennedy NP, Mohamed BM, Abuzakouk M, Dunne J, Byrne G, Feighery CF (2013) Immunological indicators of coeliac disease activity are not altered by long-term oats challenge. Clin Exp Immunol 171(3):313–318. doi:10.1111/cei.12014

8. Mohamed BM, Feighery C, Williams Y, Davies A, Kelleher D, Volkov Y, Kelly J, Abuzakouk M (2008) The use of Cellomics to study enterocytes cytoskeletal proteins in coeliac disease patients. Cent Eur J Biol 3(3):258–267

9. Zock J (2009) Applications of high content screening in life science research. Comb Chem High Throughput Screen 12(9):870–876. doi:10.2174/138620709789383277

10. Marsh MN (1992) Mucosal pathology in gluten sensitivity. In: Marsh MN (ed) Coeliac disease. Blackwell Scientific, Oxford, pp 136–191

Chapter 9

HLA Genotyping: Methods for the Identification of the HLA-DQ2,-DQ8 Heterodimers Implicated in Celiac Disease (CD) Susceptibility

Maria Edvige Fasano, Ennia Dametto, and Sandra D'Alfonso

Abstract

In this chapter we will present the principal technical methods to genotype the HLA-DQA1* and -DQB1* alleles associated with celiac disease (CD), corresponding to the serological heterodimers HLA-DQ2 and -DQ8. We will present the methods specific for the genotyping of these heterodimers, which represents a common request from consultant doctors. Because these alleles are also common in healthy subjects, their presence is not diagnostic for CD. Conversely, their absence is more important because it excludes the disease, since CD patients negative for these heterodimers are very rare. Accordingly, HLA typing has been included as a useful test to exclude celiac disease in the ESPGHAN guidelines for diagnosis of celiac disease.

The methods for HLA typing described in the present chapter are based on the following techniques:

PCR-SSP (Polymerase Chain Reaction-Sequence Specific Primers): PCR with primers specific for HLA alleles encoding the CD risk heterodimers, whose presence is revealed through the electrophoresis of PCR products.

Reverse PCR-SSOP (PCR-Sequence Specific Oligonucleotide Probes): PCR with primers specific for a single locus or a large group of alleles followed by hybridization with enzyme-conjugated probes specific for a single allele, immobilized on different supports (i.e., nitrocellulose strips), in which DNA-probes binding is revealed by the production of a colored precipitate derived from the enzymatic modification of a specific substrate.

Real-Time PCR (RT-PCR): PCR with locus or allelic specific primers whose amplification is revealed by particular probes (i.e., Taqman probes) hybridizing the DNA template within the two PCR primers and emitting fluorescent while the PCR reaction occurs.

Key words Celiac Disease, HLA-DQ heterodimers, PCR-SSP, Reverse PCR-SSOP, Real-time PCR

1 Introduction

Celiac disease (CD) is an enteropathy mediated by intolerance to gluten [1]. Although pathogenic mechanism involves both genetic and environmental factors, the involvement of genes within the human major histocompatibility complex (HLA) is well established.

Anthony W. Ryan (ed.), *Celiac Disease: Methods and Protocols*, Methods in Molecular Biology, vol. 1326,
DOI 10.1007/978-1-4939-2839-2_9, © Springer Science+Business Media New York 2015

The CD risk HLA-DQA1* and DQB1* alleles encode the alpha and beta chain, respectively, of the HLA-DQ heterodimer. Most CD patients ($\approx$90 %) are HLA-DQA1*05-DQB1*02 [2, 3] positive and hence carrier of the known high-risk serological heterodimer HLA-DQ2. The risk of CD is higher in individuals homozygous for DQB1*02 [4]. The majority of CD patients who test negative for the high-risk DQ2 heterodimer carry the HLA-DQA1*03-DQB1*0302 alleles encoding for the heterodimer serologically denoted as HLA-DQ8. The remaining CD patients (<5 %) present either DQA1*05 or DQB1*02 alone, thus carrying only half of the risk heterodimer [5]

CD patients negative for any of these HLA alleles are very rare. Therefore, the absence of both HLA-DQ2 and HLA-DQ8 heterodimer makes diagnosis of celiac disease very unlikely (sensitivity >96 %). Accordingly, HLA typing of patients has been included as a useful test to exclude celiac disease in the ESPGHAN guidelines for CD diagnosis [6]. Conversely, since these alleles are common also in healthy subjects (~40 %, ref. 5), HLA typing has very poor positive predictive value when considered for CD diagnosis [6].

The laboratory can choose to perform the test required for the celiac disease association using a complete typing, with an intermediate resolution of the whole HLA-DQA1 and HLA-DQB1 loci, or specific reagents only for the CD risk alleles reported in Fig. 1. The choice depends on the workup and the available equipment in each laboratory. For complete genotyping many methods can be chosen, some of them requiring the presence of specific apparatus, for example when Luminex technology is employed.

In the construction of primers and probes used for the typing of the CD risk heterodimers, all the necessary alleles should be covered avoiding the ambiguities between the relevant alleles for CD association risk. Figure 1 reports the minimal set of HLA alleles necessary to identify the CD risk heterodimers: they include the HLA-DQA1 and DQB1 alleles encoding for the CD risk heterodimers (indicated in bold) as well as other DQA1 and DQB1 alleles together to some HLA-DRB1 alleles not involved in the CD association but necessary to cover the different possible combinations encoding the CD risk heterodimers, in *cis* and/or in *trans*, and to detect the homozygous status for DQB1*02.

DRB1*03	DQA1*01	**DQB1*02 (02:01, 02:02)**
DRB1*04	**DQA1*03(03:01)**	DQB1*03 (03:01, 03:04)
DRB1*07	**DQA1*05(05:01, 05:05)**	**DQB1*03 (03:02**, 03:05)
DRB1*11	DQA1*06	DQB1*03:03
DRB1*12	DQA1*02(02:01)	DQB1*04

Fig. 1 HLA-DRB1*, -DQA1, and -DQB1* alleles necessary to identify the CD risk heterodimers. In *bold* are indicated the CD risk alleles; in *brackets* the alleles that must be covered

Indeed, the interpretation of the results takes advantage of the strong quite absolute linkage disequilibrium between the HLA-DQA1, -DQB1, and -DRB1 alleles [7]: this allows the attribution of predisposing haplotype with complete certitude, even in the absence of an extended complete HLA genotyping.

A typical example is represented by individuals carrying both DR7-DQ2 (DRB1*07, DQA1*02, DQB1*02) and DR11-DQ7 (DRB1*11, DQA1*05, DQB1*03) haplotypes: none of these haplotypes carries a CD risk heterodimer; however the combination of the alpha chain and beta chain in *trans* leads to the presence of the complete DQ2 heterodimer DQA1*05, DQB1*02.

The guidelines of the different Scientific Societies recommend summary of the results of each tested individual by specifically indicating the presence or absence of CD risk heterodimers, as exemplified in Fig. 2, even in the case of complete HLA genotyping extended to all HLA-DQA1 and HLA-DQB1 loci.

In this chapter we will describe the methods and reagents for the CD typing which can be part of commercial kit or of homemade procedures. If the laboratory uses homemade reagents and procedures, long and accurate validation is needed that should be repeated every time the lot number of the reagents changes. The time required for validation might be not compatible with clinical needs, although these tests cannot be considered as urgent. In general, use of commercial kits and reagents is recommended in clinical setting, because these products are validated and guaranteed by the company for diagnostic purposes (see also *see* **Note 1**).

Name of the patient   Date of the sample arrival
HLA genotyping method
Clinical motivation for the HLA genotyping

Results:
-HLA genotype:

HLA-DQA1*.................DQB1*...........................DRB1*....................

-Presence of the HLA-DQ2 heterodimer (DQA1*05-DQB1*02)   ☐

   DQB1*02 homozygote status:
   Present ☐   absent ☐   not de termined ☐

- Presence only of the beta chain of the DQ2 heterodimer (DQB1*02) ☐

- Presence of the HLA-DQ8 heterodimer (DQA1*03-DQB1*0302)   ☐

- Absence of HLA risk alleles-heterodimers   ☐

Name and Signature of the Laboratory operator and the Laboratory Director   *Date*

Fig. 2 Example of the report of the laboratory results of the HLA genotyping to detect CD risk alleles

a

```
AA Codon           55              60              65              70              75
DQA1*01:01   GGA GGT TTT GAC CCG CAG GGT GCA CTG AGA AAC ATG GCT GTG GCA AAA CAC AAC TTG AAC ATC ATG ATT AAA CGC
DQA1*02:01   A-- ... --- --- --- --A TT- --- --- -C- --- --C --- --- CT- --- --T --- --- --- --- C-- --- --- ---
DQA1*03:01   A-- A-A --- --- --- --A TT- --- --- -C- --- --C --- --- CT- --- --T --- --- --- --- G-- --- --- ---
DQA1*04:01   A-- ... --- --- --- --A TT- --- --- -C- --- --C --- --- A-- --- --- --- --- --- --- C-- --- --- ---
DQA1*05:01   A-- ... --- --- --- --A TT- --- --- -C- --- --C --- --C CT- --- --T --- --- --- --GT C-- --- --- ---
DQA1*06:01   A-- ... --- --- --- --A TT- --- --- -C- --- --C --- --- A-- --- --- --- --- --- --- C-- --- --- ---
```

b

IHWC	Cell line	DQA1*
9019	DUCAF	05:01
9098	MT14B	03:01

Fig. 3 (a): DQA1* nucleotide positions useful for the definition of HLA-DQA1 alleles associated with CD. **(b)** DQA1* genotypes of the reference cell lines carrying the CD risk alleles. Sequences have been downloaded from: http://www.ebi.ac.uk/cgi-bin/ipd/imgt/hla/align.cgi

a

```
AA Codon                      10              15              20              25
DQB1*05:01   AG GAT TTC GTG TAC CAG TTT AAG GGC CTG TGC TAC TTC ACC AAC GGG ACG GAG CGC GTG CGG GGT GTG ACC AGA
DQB1*05:02   -- --- --- --- --- --- --- --- --- --- --- --- --- --- --- --- --- --- --- --- --- --- --- --- ---
DQB1*05:03   -- --- --- --- --- --- --- --- --- --- --- --- --- --- --- --- --- --- --- --- --- --- --- --- ---
DQB1*03:01   -- --- --- --- --- --- --- --- -C- A-- --- --- --- --- --- --- --- --- --- --- --T TA- --- --- ---
DQB1*03:02   -- --- --- --- --- --- --- --- --- A-- --- --- --- --- --- --- --- --- --- --- --T CT- --- --- ---
DQB1*03:03   -- --- --- --- --- --- --- --- --- A-- --- --- --- --- --- --- --- --- --- --- --T CT- --- --- ---
DQB1*03:04   -- --- --- --- --- --- --- --- -C- A-- --- --- --- --- --- --- --- --- --- --- --T TA- --- --- ---
DQB1*03:05   -- --- --- --- --- --- --- --- --- A-- --- --- --- --- --C --- --- --- --- --- --- --- --- --- ---
DQB1*04:01   -- --- --- --- -T- --- --- --- --- A-- --- --- --- --- --- --C --- -T- --- --- --- --- --- --- ---
DQB1*06:01   -- --- --- --- CT- --- --- --- -C- A-- --- --- --T --- --- --- --- --- --- --- --T TA- --- --- ---
DQB1*02:01   -- --- --- --- --- --- --- --- --- A-- --- --- --- --- --- --A --- --- --- --T CT- --- --G- ---
DQB1*02:02   -- --- --- --- --- --- --- --- --- A-- --- --- --- --- --- --A --- --- --- --T CT- --- --G- ---
```

b

IHWC	Cell line	DQB1*
9019	DUCAF	02:01
9098	MT14B	03:02

Fig. 4 (a) DQB1* nucleotide positions useful for the definition of HLA-DQB1 alleles associated with CD. **(b)** DQB1* genotypes of the reference cell lines carrying the CD risk alleles. Sequences have been downloaded from: http://www.ebi.ac.uk/cgi-bin/ipd/imgt/hla/align.cgi

The nucleotide positions of DQA1* and DQB1* alleles suggested as crucial to identify the CD risk alleles are reported in Figs. 3, 4, and 5, together with the genotypes of reference cell lines carrying the CD risk alleles. These indications are particularly useful in case the laboratory opts for a homemade procedure. In particular, these positions are important to design PCR primers and probes specific for the CD risk alleles according to the methods described below.

The most common HLA genotyping methods employ the technique of PCR-Sequence Specific Primers (SSP), reverse PCR-Sequence Specific Oligonucleotide Probes (SSOP), and PCR-Real Time (RT).

The SSP method employs multiplex PCR (allele-specific primers plus a primer pair for an unrelated human gene, as an internal control), distributed in a variable number of tubes depending on

a

```
AA Codon      55                  60                  65                  70                  75
DQB1*05:01  CGG CCT GTT GCC GAG TAC TGG AAC AGC CAG AAG GAA GTC CTG GAG GGG GCC CGG GCG TCG GTG GAC AGG GTG TGC
DQB1*05:02  --- --- AGC --- --- --- --- --- --- --- --- --- --- --- --- --- --- --- --- --- --- --- --A --- ---
DQB1*05:03  --- --- -AC --- --- --- --- --- --- --- --- --- --- --- --- --- --- --- --- --- --- --A --- ---
DQB1*03:01  -C- --- -AC --- --- --- --- --- --- --A --- --- --- --- A-- A-- --- --- GA- T-- --- -C- --- ---
DQB1*03:02  -C- --- -CC --- --- --- --- --- --- --- --- --- --- --- A-- A-- --- --- GA- T-- --- -C- --- ---
DQB1*03:03  -C- --- -AC --- --- --- --- --- --- --- --- --- --- --- A-- A-- --- --- GA- T-- --- -C- --- ---
DQB1*03:04  -C- --- -CC --- --- --- --- --- --- --- --- --- --- --- A-- A-- --- --- GA- T-- --- -C- --- ---
DQB1*03:05  -C- --- -CC --- --- --- --- --- --- --- --- --- --- --- A-- A-- --- --- GA- T-- --- -C- --- ---
DQB1*04:01  --- -T- -AC --- --- --- --- --T --- --- --- --C A-- --- --- -A- -A- --- --- --- --- --- --CC --A ---
DQB1*06:01  --- --- -AC --- --- --- --- --- --- --- --- --C A-- --- --- A-- A-- --A --- GA- T-- --- -C- --- ---
DQB1*02:01  -T- --- -CC --- --- --- --- --- --- --- --- --C A-- --- --- A-- AAA --- --- G-- --- --- --- --- ---
DQB1*02:02  -T- --- -CC --- --- --- --- --- --- --- --- --C A-- --- --- A-- AAA --- --- G-- --- --- --- --- ---
```

b

IHWC	Cell line	DQB1*
9019	DUCAF	02:01
9098	MT14B	03:02

Fig. 5 (**a**) DQB1* nucleotide positions useful for the definition of HLA-DQB1 alleles associated with CD (**b**) DQB1* genotypes of the reference cell lines carrying the CD risk alleles. Sequences have been downloaded from: http://www.ebi.ac.uk/cgi-bin/ipd/imgt/hla/align.cgi

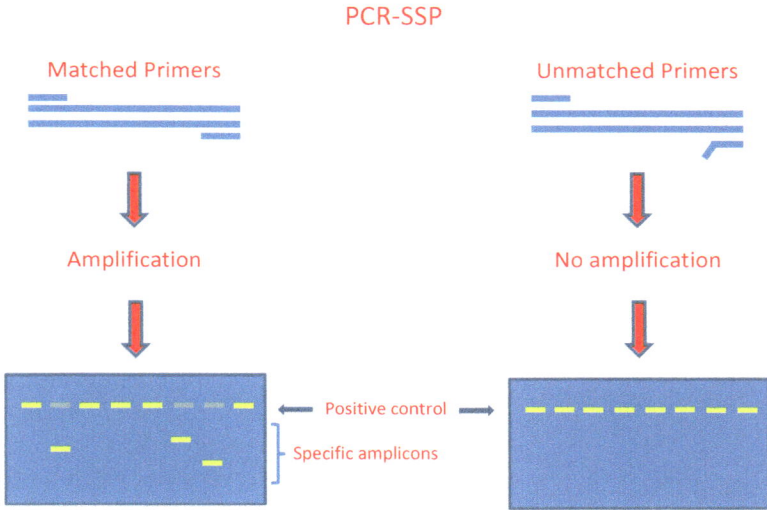

Fig. 6 Basis of the PCR-SSP method. In the PCR-SSP test, perfect match of specific primers allows amplification and detection of specific alleles

the specific number of alleles to be tested. Only primers with a perfect matching with the DNA template will allow amplification (Fig. 6). The PCR products are loaded on a gel for the electrophoresis and visualized by a UV transilluminator. For each HLA allele-specific PCR reaction, the presence of the specific PCR product indicates the positivity for that HLA allele, while its absence, in the presence of the product of the co-amplified internal control, indicates the negativity for that HLA allele. The absence of both the HLA allele-specific and the internal control product indicates a

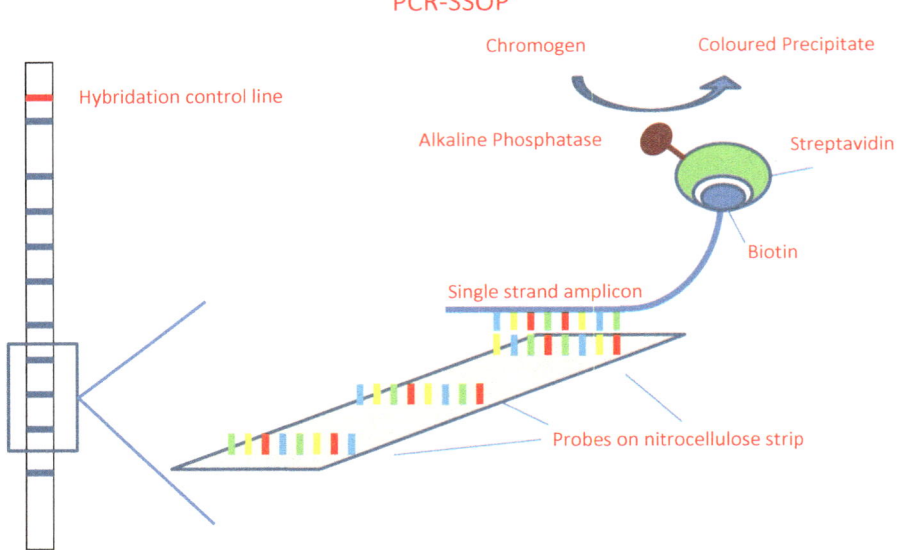

Fig. 7 Basis of the PCR-SSOP method. In the Polymerase Chain Reaction—Sequence Specific Oligonucleotide Probe (PCR-SSOP) test, perfect match between single-strand amplicon and a specific probe allows the detection of specific alleles by a colorimetric reaction

problem of the PCR reaction and hence unreliable results that need to be repeated. A photographic and storage system of the image is necessary. A manual interpretation and an interpretation by software are recommended in order to avoid clerical errors. It is also possible to personalize the report following the indication of the Scientific Societies of Immunogenetics.

In reverse SSOP, specific oligonucleotide probes are immobilized as parallel lines on nitrocellulose strips. The DNA is amplified with locus-specific biotinylated PCR primers. After the amplification the DNA is denatured and then transferred onto the strips. An exact match between probes and biotinylated amplified DNA is revealed by adding streptavidin (which binds biotin) conjugated with alkaline phosphatase together with its substrate: in the case of an exact match, the deposition of a colored product will be observed (Fig. 7), deriving from the enzymatic modification of the substrate. The interpretation of data can be performed by the operator and a scanner, using dedicated software which allows the assignment of alleles.

In RT-PCR chemistries, the detection of PCR amplification is done in the early phase of the reaction, in contrast to the traditional PCR described in the previous methods, where the reaction is analyzed after the last PCR cycle (end-point analysis). This method presents several advantages: (a) RT-PCR is more sensitive than endpoint PCR and can detect a twofold change (i.e., 10 vs. 20 copies of the template) while it is difficult on agarose gels to differentiate

RT-PCR

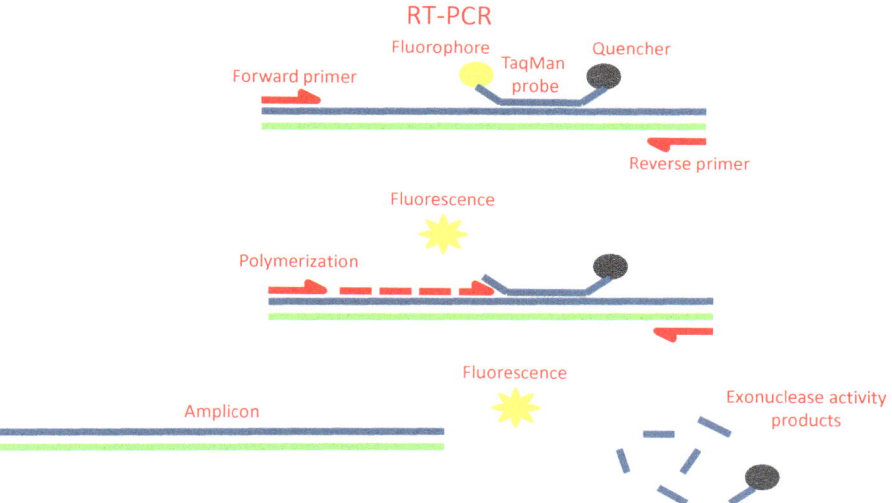

Fig. 8 RT-PCR Principle. In the Real-Time Polymerase Chain Reaction (RT-PCR) with TaqMan chemistry, a specific probe, perfectly hybridized with the sequences within the PCR amplicon, is degraded by exonuclease activity that allows fluorescence detection. The fluorescence intensity is proportional to the number of amplicons generated during the PCR and can be measured at each PCR cycle (real-time PCR)

between the fivefold changes; (b) Real Time provides fast, precise and accurate results because it is designed to collect data when the reaction is proceeding, which is more accurate for DNA quantification because it collects data during the so-called exponential phase of PCR when all the reagents are fresh and available and at every cycle the product is doubling. In subsequent cycles, some of the reagents are limiting, the reaction starts to slow down, and the PCR products are no longer doubled. (c) RT-PCR does not require laborious post-PCR methods to detect the results of the reaction. In general, the RT-PCR utilizes the so-called TaqMan™ Probes (Fig. 8), an oligonucleotide which specifically binds the single-stranded DNA internal to the PCR product between the forward and reverse primers. The probe contains a reporter fluorescent dye on the 5′ end (R, as VIC or FAM fluorophore) and a quencher dye on the 3′ end (Q), preventing the detection of the fluorescent probe. During the annealing phase of the PCR, the probe binds to the template DNA. The annealing temperature of the probe is of 5–10 °C higher than the annealing temperature of the primers, then the primers anneal to the template DNA containing the hybridized quenched probe. Thanks to its 5′-exonuclease activity, the Taq DNA Polymerase, during the elongation of a growing DNA chain, removes the downstream hybridized probe which prevents its capability to synthesize the new strand [8], by cleaving the terminal nucleotides from the 5′ of the hybridized probe. Once the probe is degraded, the quencher dye is separated from the reporter fluorescent dye allowing the fluorescent detection (Fig. 8) [9].

The increase of the fluorescent signal is proportional to the amount of the amplicon produced by a given sample and when it increases to a detectable level, it can be captured by the detection instrument and displayed by the software as an Amplification Plot. The Threshold line is the level of detection or the point at which a reaction reaches a fluorescent intensity above background and it is set in the exponential phase of the amplification for the most accurate reading. The cycle at which the sample reaches this level is called the Cycle Threshold, Ct.

2   Materials

DNA is extracted from fresh or frozen peripheral blood collected in EDTA. For DNA extraction, one of several common methods can be used (columns or beads). DNA concentration can vary from 30 to 40 ng/μl; more concentrated DNA samples can be diluted with sterile demineralized water. The volume for amplification depends on the numbers of mixes necessary for the amplification of HLA-DQA1, -DQB1, and -DRB1 alleles; generally the necessary amount can vary from 150 to 200 ng.

2.1   SSP

A single PCR reaction volume of 10 μl contains:

2.1.1   Reagents

1. Taq DNA Polymerase (0.1–0.5 U).
2. PCR Buffer: 10 mM Tris–HCl Ph 8.3–8.8, 50 mM KCl.
3. 0.5–2.5 Mm $MgCl_2$.
4. dNTP mix (dATP, dCTP, dGTP, dTTP) 0.2 mM each.
5. Deionized or distilled water until the volume.
6. 5–10 Pico mole specific primers for the amplification of the HLA-DQA1, -DQB1, and –DRB1 alleles and 1–5 Pico mole human gene as internal control (this monitors the correct mechanism of the reaction).
7. DNA 40–200 ng/reaction.

The PCR products are loaded on an agarose gel (2 % in TBE 0.5× with Ethidium Bromide or staining solutions such as "GelRed™") (see **Note 2**).

2.1.2   Lab Ware and Instruments

1. PCR tube strips (0.2 ml).
2. Adjustable volume pipettes set and related tips.
3. Disposable sterile 1.5 ml tubes.
4. Thermal Cycler.
5. Agarose gel electrophoresis equipment.
6. Photographic apparatus.
7. Interpretation sheets or software.

2.2 SSOP

2.2.1 A Single PCR Reaction Volume of 50 μl Contains

1. Taq DNA Polymerase (0.5–2.5 U).
2. PCR Buffer: 10 mM Tris–HCl Ph 8.3–8.8, 50 mM KCl.
3. 4 mM $MgCl_2$.
4. dNTP mix (dATP, dCTP, dGTP, dTTP) 0.2 mM each.
5. Deionized or distilled water until the volume.
6. 25 Pico mole Biotinylated specific primers for the amplification of the HLA-DQA1, -DQB1, and –DRB1 alleles.
7. DNA 200–500 ng/reaction

The PCR products may be loaded on an agarose gel (2 % in TBE 0.5× with Ethidium Bromide or staining solutions such as "GelRed™") (*see* **Note 2**).

2.2.2 Reagents for Detection Phase

1. Denaturing solution containing 0.4 M NaOH and 10 mM EDTA.
2. Nitrocellulose strips coated with specific oligonucleotides.
3. 20× SSPE (4.5 M NaCl, 0.3 M NaH_2PO_4, 30 Mm EDTA. The solution must be 7.4Ph).
4. SDS 20 %.
5. Hybridization Buffer: SSPE4× and 0.5 % SDS for the binding between amplification product and probes.
6. Wash Stringent Buffer: 1× SSPE and 0.1 % SDS.
7. Ambient Wash Buffer: 1× SSPE and 0.1 % SDS.
8. Conjugate solution, containing streptavidin conjugated with alkaline phosphatase (commercial).
9. Color developer solution, containing alkaline phosphatase substrates such as 5-Bromo-4-chloro-3-indolyl-phosphate and 4-Nitroblue tetrazolium.

2.2.3 Lab Ware and Instruments

1. Adjustable volume pipettes set and related tips.
2. Disposable sterile 1.5 ml tubes.
3. Thermal Cycler.
4. Agarose gel electrophoresis equipment.
5. Photographic apparatus.
6. Plastic tray.
7. Shaking water bath with adjustable temperature.
8. Vacuum aspiration apparatus.
9. Interpretation sheets or software.

2.3 RT-PCR

2.3.1 Reagents for Single Reaction

1. Different mixes of lyophilized primers and probes (TaqMan chemistry) to detect specific HLA-DQA1 and -DQB1 alleles and an internal control as β-globin human gene (this increases the accuracy of the reaction). Two reporters, such as VIC or FAM, are used to distinguish different products of reaction.

2. Buffer TE (Tris/EDTA) 1× for the reconstitution of primers and probes.

3. Master Mix, usually ready to use with Taq DNA Polymerase for PCR Real Time (AmpliTaq Gold® DNA Polymerase) and dNTPs.

4. DNA 50–70 ng.

2.3.2 Lab Ware and Instruments

1. Adjustable volume pipettes set and related tips.

2. Disposable sterile 1.5 ml tubes.

3. PCR Real-Time thermal cycler.

4. Export file interpretation software.

3   Methods

3.1   SSP

1. PCR mix is prepared adding master mix, water, Taq, and DNA, and then it is dispensed into micro tubes containing specific primers. Before DNA is added to the master mix, an aliquot is dispensed into the negative control tube if required (*see* **Note 3**).

2. Thermal cycler is set according to the profile that can change depending on the primers employed (up to 30 cycles in which annealing temperature varies from 70 to 55 °C).

3. The amplification products are loaded on an electrophoretic gel and a picture can be taken with a camera. All internal controls must be present in each well, with the exception of the negative control. Specific amplifications indicate positive results.

4. The operator checks the positive and negative mixes and assigns a result using the interpretation sheet or software provided.

3.2   SSOP

3.2.1   Amplification

1. PCR mix is prepared adding master mix, primers, water, Taq, and DNA. For each work session, one tube must be provided to control contamination, where all components are added except DNA.

2. Reaction tubes are placed in the thermal cycler where an appropriate amplification program is set (up to 35 cycles in which annealing temperature varies from 62 to 64 °C, depending on the primers used).

3. The amplification products may be checked on agarose gel (*see* **Note 4(a)**).

3.2.2   Detection

Detection can be performed using an automatic instrument or a manual procedure. Setup of instrument can be made following specific procedure, whereas in case of manual detection these steps must be followed.

1. DNA denaturation

   PCR products of each sample are mixed with denaturing solution and incubated at room temperature.

2. Hybridization

   (a) One strip must be removed for each sample using clean tweezers (to be touched only with gloves) and marked with a pencil (do not use ballpoint pen, markers, etc.).

   (b) Each strip is placed into a lane in a plastic tray with marker lines facing up and completely soaked.

   (c) Some microliters of denatured DNA must be put into each respective lane and 1 ml of preheated Hybridization Buffer (*see* **Note 4(b)**).

   (d) Incubation (45 to 50 °C) must be carried out for an appropriate time (from 15 to 30 min) in the shaking water bath (about 50 rpm) set; the water bath should be closed to avoid temperature variations (*see* **Note 4(c)**).

3. Wash.

   (a) This should be carried out at the same temperature as the hybridization with shaking.

   (b) When the hybridization is completed, the operator removes the tray from the water bath and aspirates the solution from each lane using a pipette or a vacuum aspiration apparatus, without touching the strip surface.

   (c) After a short rinse with the pre-warmed wash stringent solution (*see* **Note 4(b)**), the operator repeats two washes with the same reagent at the same hybridization temperature for 10 min.

4. Binding detecting and color development

   (a) This should be carried out at room temperature, with shaking.

   (b) After the wash steps, an incubation (20 min) with an appropriate volume of conjugate solution must be performed.

   (c) The operator removes the conjugate solution and rinses briefly twice with fresh wash solution (room temperature), and then incubates (5 min) twice with the same solution (*see* **Note 4(d)**).

   (d) The operator removes the wash solution and adds the color developer solution. The incubation (20 min) must be performed in the dark.

   (e) The operator removes the color developer solution and performs wash with distilled water.

   (f) Before reading, strips must be dried in the dark.

(g) The operator checks the positive bands and assigns a result using the interpretation sheet.

(h) Some companies provide the laboratory with a scanner and software for the reading and the interpretation (*see* **Note 4(e)**).

3.3 RT-PCR

3.3.1 Setting Up of the PCR Reaction (See Note 5)

1. Lyophilized primers and probes are reconstituted with Buffer TE 1×. 5 μl of the mix must be transferred in tube strips (0.2 ml) in a 96 well micro plate. After deposition the tubes must be closed and centrifuged quickly.

2. Master Mix must be carefully mixed and centrifuged before use.

3. 15 μl of Master Mix must be put in the negative control tube of the reaction or No Template Control (NTC). It is advisable to close this tube before proceeding.

4. 750–1000 ng of DNA must be added to 330 μl of Master Mix: mix carefully and centrifuge.

5. 20 μl of this mixture can be dispensed in each position of the micro plates where the primers and probes reconstituted have been transferred. The micro plate can be closed and centrifuged.

6. The procedure must be repeated for each sample until the micro plate is completed with the samples needed.

3.3.2 Real-Time Amplification

1. The procedures for the setting up of the RT-PCR are different depending on the instrument used; usually, the thermal cycler must be set for one denaturing cycle (10 min 95 °C) and 37 amplifying cycles (15 s 95 °C, 1 min 30 s 60 °C).

2. A new document must be created for each work session, in which are set the fluorophores (VIC and/or FAM) used in each position of the plate.

3. The reading of the instrument must be set at step at 60 °C and the volume of reaction at 25/μl.

3.3.3 Result Validation

1. The results visualized can be modified and visualized for each fluorophore and position of the plate.

2. The operator must check that in the position of the NTC, with the corresponding reading filter (e.g., VIC), the reaction is ≥ value 34. When the Ct is lower, a contamination of the reagents used is possible, the results are not reliable, and the test must be repeated.

3. The mixes containing specific primers for HLA loci must show a signal with the corresponding VIC filter (e.g., Ct between 21 and 28) when positive. Ct values below 21 are not acceptable and the reaction should be repeated.

4. Position after position, the reading with specific filter (e.g., FAM) must proceed. Values ≥ 18 indicate positive reaction.

5. Then the analysis of the results can be performed.

4   Notes

1. Choice of the Genotyping Method

   The laboratory should have at least two methods for typing CD alleles, a principal which has been employed routinely and another of support in case of not satisfactory or confirmatory results. The SSP method is particularly suitable when the laboratory needs to type a few samples and in the case of confirmatory results because in a short time and with a moderate workup the results can be achieved. The PCR-SSOP is optimal for handling numerous samples at once. On the other hand, RT-PCR offers some guaranties of quality and specificities due to the presence of primers and probes on the same reaction. See the text for details on these techniques.

2. Reagents Used for the Electrophoresis of Amplification Products

   Ethidium Bromide is dangerous and must be handled with caution. Gels should be prepared under a hood when Ethidium Bromide is added to agarose gels. Gel Red can be used instead of Ethidium Bromide because it is supposed to be less dangerous but the laboratory must compare with accuracy the results obtained before introducing it in the routine practice.

3. SSP

   In case of DNA concentration lower than the reference of the test, the laboratory can decide to repeat the DNA extraction or to perform as well the test using a major volume of DNA of some microliters. When the DNA purity is not as good, and 260/280 is <1.8, the DNA extraction can be repeated or greater quantity of Taq polymerase can be used.

4. SSOP

   (a) Verification of the amplicons: In the European Federation of Immunogenetics (EFI) Standards (last version 6.1) in the Section L (11.3.1) it is recommended to monitor the amplification before hybridization. In case of numerous amplified samples, the workup for checking the amplification on a gel can be long and prone to technical mistakes. In this case, the laboratory can decide to control randomly only some of the amplified samples.

   (b) The Hybridization and Stringent Wash Solution must be pre-warmed and all crystals should be completely dissolved.

   (c) The water bath must reach the temperature indicated by the producer of the reagent (from 45 °C to 56 °C) and the operator checks using a calibrated thermometer (tolerance is permitted of 0.5 °C). The water level of the shaking bath should be adjusted to approx. 2/3 of the height of the Plastic Tray.

(d) Wash Ambient Solution, Conjugate and Color developer must reach room temperature.

(e) Use of the scanner

When the laboratory uses the scanner for the interpretation of the strips, it can happen that some spots are very pale and the machine is not able to count them. In this case, it can be useful that the operator checks the spot manually and makes an appropriate interpretation.

5. RT-PCR

The operator must be very careful in the preparation of the mix. Two pairs of gloves, one on the other, must be used. The gloves must be changed when there is a suspicion of contamination. To avoid contamination, it is advisable to work under a laminar flow hood or use a facemask.

References

1. Tries JS (1999) Coeliac sprue. N Engl J Med 325:1709–1719

2. Sollid LM, Markussen G, Ek J et al (1989) Evidence for a primary association of coeliac disease to a particular HLA-DQα/β heterodimer explains the divergent HLA-DR associations observed in various Caucasian populations. J Exp Med 169:345–350

3. Sollid LM, Thorsby E (1990) The primary association of coeliac disease to a given HLA-DQα/β heterodimer explains the divergent HLA-DR association observed in various Caucasian populations. Tissue Antigens 36:136–137

4. Veoden W, Stepniak D, Kooy Y et al (2003) The HLA-DQ2 genes dose effect in coeliac disease is directly related to the magnitude and breadth of gluten-specific T cell responses. Proc Natl Acad Sci USA 100:12300–12305

5. Margaritte-Jeannin P, Babron MC, Bourgey M, Louka AS, Clot F, Percopo S, Coto I, Hugot JP, Ascher H, Sollid LM, Greco L, Clerget-Darpoux F (2004) HLA-DQ relative risks for coeliac disease in European populations: a study of the European Genetics Cluster on Coeliac Disease. Tissue Antigens 63:562–567

6. Husby S, Koletzko S, Korponay-Szabo´ R, Mearin ML, Phillips A, Shamir R, Troncone R, Giersiepen K, Branski D, Catassi C, Lelgeman M, Maki M, Ribes-Koninck C, Ventura A, Zimmer KP, for the ESPGHAN Working Group (2012) European Society for Pediatric Gastroenterology, Hepatology, and Nutrition Guidelines for the Diagnosis of Coeliac Disease. JPGN 54:136–160

7. Begovich AB, McClure GR, Suraj VC et al (1992) Polymorphism, recombination, and linkage disequilibrium within the HLA Class II Region. J Immunol 148:249–258

8. Holland PM, Abramson RD, Watson R et al (1991) Detection of specific polymerase chain reaction product by utilising the 5′ to 3′ exonuclease activity of Thermus aquaticus DNA polymerase. Proc Natl Acad Sci U S A 88:7276–7280

9. Livak KJ, Flood J, Marmaro W et al (1995) Oligonucleotides with fluorescent dyes at opposite ends provide a quenched probe system useful for detecting PCR product and nucleic acid hybridization. PCR Methods Appl 4:357–362

Detecting Allelic Expression Imbalance at Candidate Genes Using 5′ Exonuclease Genotyping Technology

Jillian M. Gahan, Mikaela M. Byrne, Matthew Hill, Emma M. Quinn, Ross T. Murphy, Richard J.L. Anney, and Anthony W. Ryan

Abstract

Genetic variation along the length of a chromosome can influence the transcription of a gene. In a heterozygous individual, this may lead to one chromosome producing different levels of RNA, compared to its paired chromosome, for a given gene. Allelic differences in gene expression can offer insight into the role of variation in transcription, and subsequently infer a route to conferring disease risk. This phenomenon is known as allele expression imbalance or AEI, which may be assayed using a PCR-based method that includes the quantification of the relative dosage of each allele (e.g., 5′ exonuclease assays, TaqMan™). Importantly, in heterozygous individuals the resolution of expression imbalance is performed within a controlled system; the comparison of the alternate allele is reported relative to the wild-type, as the experiment can be performed within a single sample, controlled for background genetic information. Alternative methods for the detection of AEI include Primer-extension MALDI-TOF (Sequenom MassARRAY®), Next-Generation Sequencing, and SNP genotyping arrays. Here we present the methods used for the TaqMan™ approach and include a description of the SNP identification, allele-specific PCR, and analytic methods to convert allele amplification metrics to relative allele dosage.

Key words Allelic expression imbalance, 5′ exonuclease assay, cDNA, CEPH cell lines

1 Introduction

Gene expression can be influenced by *cis*- and *trans*-genetic variation. In an individual who is heterozygous at a genetic variant which influences gene expression, the gene exhibits allelic expression imbalance (AEI), whereby one allele produces more mRNA than the other. This phenomenon is widespread throughout the genome and may be detected using allele-specific quantitation assays such as oligonucleotide array technology [1], transcriptome

Electronic supplementary material The online version of this chapter (doi:10.1007/978-1-4939-2839-2_10) contains supplementary material, which is available to authorized users.

Anthony W. Ryan (ed.), *Celiac Disease: Methods and Protocols*, Methods in Molecular Biology, vol. 1326, DOI 10.1007/978-1-4939-2839-2_10, © Springer Science+Business Media New York 2015

sequencing (RNA-seq) [2], or 5′ exonuclease assays [3]. Analysis of AEI provides a powerful method to detect *cis*-acting genetic variants and may uncover a functional role for noncoding SNPs in GWAS analyses. AEI has primarily been a candidate gene driven application. However, genome-wide AEI can be evaluated using genomic technologies. For example, RNA-seq analysis—a transcriptome-wide next-generation sequencing approach—has been applied to primary cells of human subjects [2]. This has the advantage of detecting imbalance across the entire transcriptome in a single experiment. However, it is not without pitfalls—the strength of the methodology is in the ability to accurately estimate the allele dosage; for RNA-seq, the ability to reliably detect AEI depends on the depth of RNA-seq coverage at any region of the genome—if this is too low due to low abundance, the ability to detect either allele confidently may be compromised. In this chapter, we highlight the necessary steps and the potential pitfalls in performing AEI, focusing on the TaqMan™ 5′ exonuclease assay for allele-specific amplification.

The basic concepts of AEI are shown in Fig. 1. Briefly, the paternal (blue) and maternal (pink) chromosome are depicted. The noncoding (lowercase) sequence contains a single purine polymorphism (g/a) which influences the "brightness" of expression of the gene (uppercase). The result is a biased over transcription from the paternal chromosome. The exon (uppercase) possesses a second a/t variant that can be used for allele-specific amplification in both the genomic and cDNA. Upon allele-specific amplification using the TaqMan™ 5′exonuclease method, the overexpressed allele amplifies more rapidly, and the Ct (threshold cycle)—the cycle at which the relative fluorescence intensity (Rn) for each allele is detected above a threshold—is lower. That is, the paternal allele, with its higher starting copy number, requires fewer amplification cycles to achieve a given amplification threshold.

Detection of AEI using Taqman™ technology requires the identification of a SNP, which is present on the mature, spliced transcript. The variant must be located sufficiently far from splice

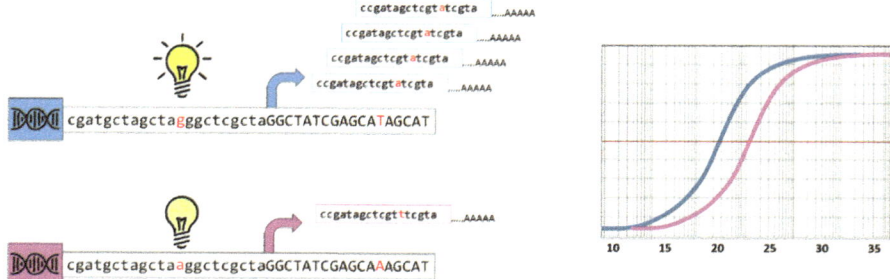

Fig. 1 Cartoon depicting the allele expression imbalance model

junctions to allow both cDNA and genomic DNA PCR-based assays to function. Where this is not possible, researchers have attempted to measure transcriptional activity using DNA fragments purified from chromatin immunoprecipitation assays of RNA polymerase II binding [4]. Taqman™ detection of AEI requires amplification of matched genomic DNA and cDNA samples to account for any spurious assay biases. For example, allele-specific differences may arise if the binding of the allele-specific probe is compromised by a second variant in linkage disequilibrium with the test allele (e.g., a second SNP that resides within the binding sequence), or where the secondary structure of the resultant PCR amplicon influences the amplification efficiency in an allele-specific manner. Additionally, analyses of the genomic DNA can identify dosage differences (or copy number variation) which will in turn impact on the cDNA copy number. If we assess the assay in the genomic DNA and adjust for any biases, we can reduce these technical biases and have a greater confidence that the differences observed result from the biology of the test variant. Importantly, because AEI is relative to the reference allele, unlike other quantitative PCR methods there is no need to determine the actual quantity of input material. A typical example of AEI quantification is shown in Fig. 2. It is important to note that the gDNA traces overlay to indicate equal dosages, whereas the cDNA traces are separated by approximately a single amplification cycle.

At a single locus, AEI is evaluated by performing 5′exonuclease (Taqman™) genotyping on the genomic DNA and the cDNA [3]. Relative levels of allele dosage may be calculated using the ΔCt method [5]. The statistical procedure for differentiating output data to relative allele dose was developed from DNA pooling applications [6]. Threshold amplification cycle data (Ct: the cycle at which the relative fluorescence intensity (Rn) for each allele is detected above a threshold) are used to determine relative allele frequency (rF) of allele 1 compared to allele 2. In essence, a single cycle decrease in Ct equates to a doubling in the dosage of that allele. The equations to determine rF are given below;

$$rF = \frac{1}{2^{\Delta Ct'} + 1} \tag{1}$$

$$\Delta Ct' = \Delta Ct^{cDNA} - \Delta Ct^{gDNA} \tag{2}$$

$$\Delta Ct^{cDNA \vee gDNA} = Ct^{ALLELE1} - Ct^{ALLELE2} \tag{3}$$

rF is a relative frequency and therefore requires observation of both alleles. It is reported within a range from >0 to <1. The interpretation of the rF is that a value of 0.5 indicates that there is equal dosage of both allele 1 and allele 2 in cDNA, and an rF of 0.667 indicates a 2:1 ratio of allele 1. If only 1 allele is observed, this is the most extreme case and equates to complete allele drop.

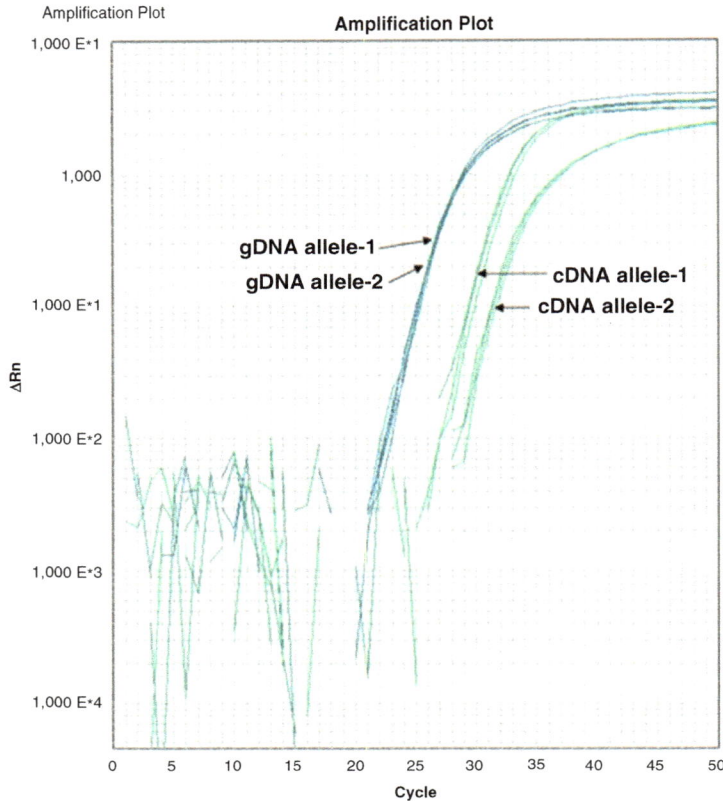

Fig. 2 Typical example of AEI traces from real-time quantitative PCR

AEI is typically considered to be present if rF deviates from 0.5 by at least 0.1; therefore AEI exists if rF < 0.4 or rF >0.6.

It is important to recognize that the tested polymorphism need not necessarily exert any functional effects itself—it is merely used to detect the imbalance due to linked variation. In Fig. 3, we simulate some examples of AEI visualized in a standard dot plot. In Panel A, we mimic an example of no AEI; all rF are reported within the 10 % boundary around 0.5. In Panel B we mimic an example of AEI in complete linkage disequilibrium with the effective variant; all rF are observed above the 0.6 boundary suggesting an overexpression of the reference allele. In Panel C and D we mimic examples where there is evidence of AEI. However, there is evidence of recombination between the causative and measured variants, as highlighted by the rF showing AEI with dosages in both directions. Additionally Panel D reveals incomplete linkage disequilibrium with the causative variant, with some markers showing no evidence of AEI. Because recombination events may occur between the causative and assayed variant, it is important not to consider the average AEI across all of your samples as a metric.

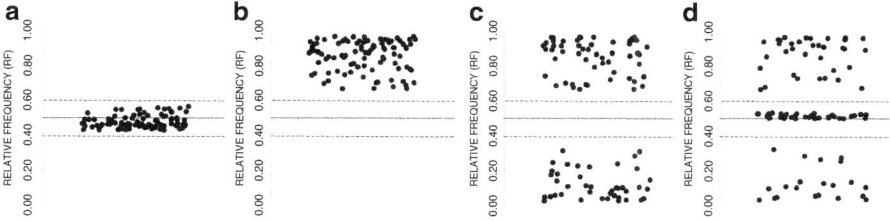

Fig. 3 Simulated example dot plots of relative frequency (rF). Central lines show the AEI thresholds 10 % limits (>0.4 to <0.6)

In examples such as Panel C this would likely infer no evidence for AEI. In such cases when it is necessary to examine many thousands of markers simultaneously, we recommend using the deviation of rF from the 50:50 ratio, such that an rF of 0.677 or 0.333 is treated equally as a 2:1 ratio. Using additional genotype data, we can also examine the AEI at the tested SNP against non-tested SNPs. In this case we can explore the Kappa-coefficient, previously applied by Lim and colleagues [7]. This examined the correlation between AEI and heterozygosity at an adjacent marker not in linkage equilibrium. Thus, $\kappa = 1$ indicates all heterozygous individuals show AEI (defined as rF < 0.4 or rF >0.6) whereas all homozygous individuals do not; conversely $\kappa = 0$ is the opposite. This is a useful tool to identify which variant is driving the AEI signal.

AEI may be explored using lymphoblastoid cell lines from reference population samples such as the *Centre d'Etude du Polymorphisme Humain* HapMap collection [3]. Using the extensive genotype data for this collection, we are able to preselect SNPs with sufficient heterozygosity. Heterozygosity and therefore allele frequency is a major limitation in performing AEI; based on Hardy-Weinberg Equilibrium, for a SNP with a 1 % minor allele frequency (MAF), less than 2 % of the sample will be heterozygous. For a SNP with MAF of 10 % and 20 %, the expected heterozygosity becomes a more opportune 18 % and 32 % of the sample respectively.

Lymphoblastoid cell lines from the CEPH HapMap collection, although convenient, will inform primarily on AEI in that cell-type alone. Although there may be correlation between tissues [8], AEI is influenced both spatially (amongst different cell types) and also temporally (at different stages of development). This is not unexpected, given our knowledge of differential gene regulation. This is particularly important when using RNA derived from whole organs. These tissues are often comprised of multiple cell populations, each expressing a unique set of regulatory factors. However, the use of bulk tissue has shown that the effect of *cis*-acting regulatory variation differs between regions of the adult human brain [9]. In addition to spatial differences, the expression of developmentally restricted regulatory factors can also influence the effect of regulatory variants on gene expression [10]. Ultimately, AEI differences are likely to extend to individual cell types, but interrogation at this level will require advances in cell sorting/isolation and preparation of RNA from limited cell numbers. In this chapter we present the methods used for the TaqMan™ approach and include description of the SNP identification, allele-specific PCR, and analytic methods to convert allele amplification metrics to relative allele dosage.

2   Materials

2.1 Selection of Candidate Single Nucleotide Polymorphism

1. Computer.
2. Internet connection.

2.2 5′ Exonuclease Assay

1. 5′exonuclease probe based assay (e.g., Taqman®).
2. 2× Master Mix.
3. 1.5 ml microcentrifuge tubes.
4. 96 well or 384 well optical plastic reaction plates.
5. Adhesive optical plate seals.

3   Methods

3.1 Selection of Candidate Single Nucleotide Polymorphism

To identify SNPs of interest, they must meet a number of criteria. Firstly they must be present at a sufficient frequency in your sample. Secondly, they must be within the gene transcript, and thirdly they must be assayable within a single exon. The assay must not cross the intron/exon splice site as it cannot then be used to assay both gDNA and cDNA. In the following section we will explain how to identify appropriate SNPs using the UCSC genome browser.

1. Navigate using a web browser to the Genome Browser Gateway (http://genome-euro.ucsc.edu/cgi-bin/hgGateway).
2. Using the *Search* facility, select the gene of interest, e.g., *LPP*.
3. Select the location based on RefGene co-ordinates.
4. In the browser, click the link for the *Common SNPs (build)* tacks under the variation tab.
5. Select limits for this track:
   (a) Display = Full
   (b) Minor Allele Frequency range = 0.2–0.5
   (c) Class = Single Nucleotide Polymorphism

(d) Validation = By HapMap; By 1000 Genomes Project

(e) Function = synonymous_variant;   nc_transcript_variant; stop_gained; missense_variant; stop_lost; frameshift_variant; 3_prime_UTR_variant; 5_prime_UTR_variant

6. From the browser, click links to the SNPs that meet these criteria (e.g., rs6768694)—you may need to adjust the parameters such as MAF to increase likelihood of identifying assayable SNPs.

7. Appropriate SNPs may be selected on the vendor's web site, for example Applied Biosystems (*see* **Note 1**). Where possible, confirm the location of the amplification probes from the assay provider.

3.2   Running the 5′ Exonuclease Assay

The 5′ exonuclease assay can be used to perform an allele-specific PCR. The principle requires two amplification primers and an allele-specific probe containing a quenched fluorophore. The probes for each allele are tagged with discrete distinguishable fluorophores. During the PCR amplification process, the fluorophore is removed from the probe and its quencher, as a result of the 5′- 3′ exonuclease activity of Taq polymerase. The free fluorophore is then detected and the accumulation of fluorescence is plotted against time (or cycle) (*see* Fig. 2).

For each AEI experiment we perform allele-specific amplification in triplicate on matched genomic DNA and in triplicate on complementary DNA. Amplification is only performed on heterozygous individuals for the test SNP as we are looking for the relative abundance of each allele and not the overall abundance of the product. The input DNA should be sourced from a heterozygous individual where we are able to extract both RNA (to generate cDNA) and matching genomic DNA. We have successfully applied this methodology to biopsied tissue, cultured cell lines, primary cells, and peripheral blood. As a result of the temporal and spatial nature of AEI, where possible we recommend using disease related and developmental stage appropriate cell types. As a rule of thumb, at least 5 ng of template DNA is required per PCR, more so for rare transcripts. If you intend to use cell lines, we recommend culturing a minimum of 10^4 cells in a 75 cm^2 flask. RNA and DNA extraction should be performed once the cells reach approximately 80 % confluence. In cultured cells this yields up to 5–7 μg DNA and 5–15 μg of RNA per 10^6 cells. We have not presented a specific gDNA, RNA, and cDNA protocol in this chapter as there are a number of published protocols and commercially available kits to enable high quality RNA and DNA extraction as well as cDNA preparation from diverse tissue samples.

1. Perform reactions in triplicate for cDNA alongside the corresponding genomic DNA (gDNA).

2. For a reaction volume of 10 μl, add 5 μl TaqMan Universal PCR Master Mix, 0.25 μl 40× TaqMan® assay mix (containing primers and allele-specific probes), and 20 ng gDNA or 1 μl of cDNA from the RT reaction.

3. When all samples have been loaded (*see* **Note 2**), cover the plate with an optical adhesive.

4. On the computer attached to the Taqman machine, open the SDS software, select File > New and "absolute quantification, 384 or 96 well."

5. On the left-hand panel, which represents the 96 or 384 well layout, highlight the area of the plate that contains your samples. Unused wells may be left empty.

6. Select "add detectors" (*see* **Note 3**), select appropriate fluorophores (e.g., FAM™-NFQ) from the list, and copy to the plate document. Then add the second fluorophore (e.g., VIC-NFQ™) in the same manner. Once both have been copied, select Done. Click on the boxes marked "use" for both markers.

7. Select the instrument tab and change the reaction volume to the appropriate value.

8. Connect to the machine by selecting "Connect." Open the door and place the plate in the Taqman machine, paying attention to the correct orientation, and close the machine using the instrument control tab.

9. Select Start, and save the file with an appropriate filename.

10. Once the run has been completed, export the Ct values to a text file. These will be used to calculate the expression of each cDNA allele relative to its genomic DNA equivalent (*see* **Note 4**).

3.3 Preparation of Biospecimens

Prior to performing the 5′exonuclease assay on patient sample, we recommend optimizing the assay conditions using control genomic DNA and control cDNA. A robust assay should render a Ct at approximately 30 cycles. Samples that show late amplification (i.e., $Ct > 35$) are less robust to replication, and it is more difficult to determine Ct than for those that amplify earlier. This may introduce spurious data points into the experiment. It may not be uncommon, for rare transcripts, to need to optimize the amount of input cDNA included in the PCR. Increasing the concentration of the input material will improve the amplification profile. As the doubling of the DNA template will reduce the Ct by approximately 1 cycle, consequently an approximate reduction of 5 cycles requires a 2^5 (or 32-fold) increase in DNA template. Researchers should not rely on summary tables of Ct outputs; all amplification curves should be manually examined for quality, with acceptable amplification curves revealing a smooth exponential growth of PCR product. Samples that show poor amplification should be discarded.

3.4 Calculation of Allelic Expression Imbalance

1. Threshold cycles (Ct) for both gDNA and cDNA are determined using the SDS software (Applied Biosystems)

2. The relative frequency or AEI is calculated using (see above for derivation):

$$rF = \frac{1}{2^{\Delta Ct} + 1}$$

3. An AEI result of 0.5 indicates a 50:50 ratio of alleles in the cDNA sample, i.e., no evidence of AEI.

4. Differences of >20 % (i.e., $0.4 > rF > 0.6$) may be considered evidence of AEI.

5. A standard dot plot is used to visualize the evidence for AEI across a population of individuals. An excel worksheet is included in the supplementary material to assist in AEI quantification (*see* supplementary worksheet 1.xlsx). An example of the dot plots produced by this worksheet is given in Fig. 4.

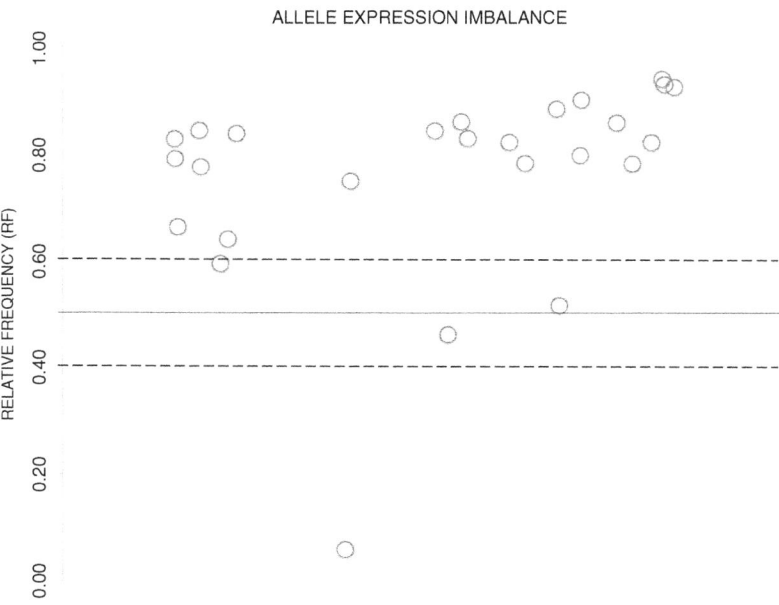

Fig. 4 Example Dot Plot showing evidence for AEI

4 Notes

1. The selected SNP must be present on the mature transcript (exonic, 5′UTR or 3′UTR). The position of the SNP should be sufficiently distant from splice junctions to ensure that the primer pair will successfully amplify cDNA. This will fail if one or both of the primers span a splice junction or are otherwise complementary to a sequence that is modified during splicing.

2. 96 Well plates may be loaded using an 8 or 12 channel pipette. Loading a 384 well plate is more difficult but can still be achieved using an 8 or 12 multichannel pipette, where pipette tips are separated by a distance of 2 wells. Wells on a 384 well plate must therefore be loaded alternatively (e.g., wells A, C, E, G etc. first, followed by wells B, D, F, H etc.) if using a multichannel pipette.

3. Fluorescent dyes used in the Taqman™ assays typically incorporate fluorophores FAM™ (6-carboxyfluorescein) and VIC™ dyes together with a nonfluorescent quencher (NFQ). It is important to confirm the quencher and also the fluorophores attributed to each allele prior to running the assay. The real-time PCR machine needs to be calibrated and programmed to detect the correct fluorophores, which may be assay and supplier specific. Details of the correct detectors for each commercially available assay are available from the supplier.

4. The Ct values should ideally be about 20–30—very high values (>35), corresponding to very low cDNA/mRNA concentrations, cannot be considered reliable.

Acknowledgements

The authors gratefully acknowledge funding from the Royal City of Dublin Hospital Trust

References

1. Pant PVK, Tao H, Beilharz EJ et al (2006) Analysis of allelic differential expression in human white blood cells. Genome Res 16:331–339

2. Heap GA, Yang JHM, Downes K et al (2010) Genome-wide analysis of allelic expression imbalance in human primary cells by high-throughput transcriptome resequencing. Hum Mol Genet 19:122–134

3. Quinn EM, Hill M, Anney R et al (2010) Evidence for cis-acting regulation of ANK3 and CACNA1C gene expression. Bipolar Disord 12:440–445

4. Knight JC, Keating BJ, Rockett KA, Kwiatkowski DP (2003) In vivo characterization of regulatory polymorphisms by allele-specific quantification of RNA polymerase loading. Nat Genet 33:469–475

5. Livak KJ, Schmittgen TD (2001) Analysis of relative gene expression data using real-time quantitative PCR and the 2-[Delta][Delta] CT method. Methods 25:402–408

6. Sham P, Bader JS, Craig I et al (2002) DNA Pooling: a tool for large-scale association studies. Nat Rev Genet 3:862–871

7. Lim J-E, Pinsonneault J, Sadee W, Saffen D (2007) Tryptophan hydroxylase 2 (TPH2) haplotypes predict levels of TPH2 mRNA expression in human pons. Mol Psychiatry 12:491–501

8. Nica AC, Parts L, Glass D et al (2011) The architecture of gene regulatory variation across multiple human tissues: the MuTHER study. PLoS Genet 7, e1002003

9. Buonocore F, Hill MJ, Campbell CD et al (2010) Effects of cis-regulatory variation differ across regions of the adult human brain. Hum Mol Genet 19:4490–4496

10. Hill MJ, Bray NJ (2012) Evidence that schizophrenia risk variation in the ZNF804A gene exerts its effects during fetal brain development. Am J Psychiatry 169:1301–1308

.

Chapter 11

Gene Expression Profiling of Celiac Biopsies and Peripheral Blood Monocytes Using Taqman Assays

Martina Galatola, Renata Auricchio, and Luigi Greco

Abstract

Quantitative real-time PCR (qPCR) allows for highly sensitive, rapid, and reproducible quantification of mRNA: it has become an established technology for the quantification of gene expression with the $5'$ nuclease assay using TaqMan® probes. It is used for a broad range of applications, including quantification of gene expression, measuring RNA interference, biomarker discovery, pathogen detection, and drug target validation. When studying gene expression with qPCR, scientists usually investigate changes—increases or decreases—in the quantity of particular gene products or a set of gene products. Investigations typically evaluate gene response to biological conditions such as disease states, exposure to pathogens or chemical compounds, organ or tissue location, and cell cycle or differentiation status. Here we describe this technique applied to molecular profiling of candidate genes in celiac biopsies and peripheral blood monocytes. Using data obtained by gene expression experiments, a discriminant equation has been developed that allows the correct classification of Celiac Disease (CD) patients compared to healthy controls, CD patients on a Gluten Free Diet (GFD), and other disease controls.

Key words Taqman assay, Gene expression, Celiac disease, Intestinal mucosa, Monocytes, Stepwise discriminant analysis

1 Introduction

Measurement of gene expression is becoming increasingly important in the study of diverse biological processes and understanding of disease pathogenesis [1]. Traditional methods, such as Northern blots and RNA protection assays, are limited by their requirement for large amounts of RNA and their time-consuming nature. By contrast, real-time reverse transcription-polymerase chain reaction (RT-PCR) requires minute amounts of RNA and is rapid and quantitative [2]. These assays require high reproducibility and precision, which may be impaired by inconsistencies in the procedures used to collect tissues and to isolate the RNA.

A common method to minimize these errors is to standardize the RNA or cDNA input amount and to simultaneously measure

Anthony W. Ryan (ed.), *Celiac Disease: Methods and Protocols*, Methods in Molecular Biology, vol. 1326,
DOI 10.1007/978-1-4939-2839-2_11, © Springer Science+Business Media New York 2015

an RNA whose expression level is constant among samples [3, 4]. This RNA serves as an endogenous control and allows comparison of the data of the genes of interest among different samples. The critical steps which must be considered in a qPCR assay for a gene expression study are as follows:

- Sample Acquisition, Handling, and Preparation: these are the first potential source of experimental variability, especially for experiments targeting RNA, because mRNA profiles are easily perturbed by sample collection and processing methods. Nucleic acid extraction is a second critical step; the efficiency depends on adequate homogenization, the type of sample, target density, physiological status, genetic complexity, and the amount of biomass processed. These are all features which must be optimized to perform a reliable experiment.

- Quantification of RNA in the extracted samples is important, because it is mandatory that approximately the same amounts of RNA be used when comparing different samples.

- Reverse transcription: it is mandatory to standardize the amount of RNA reverse-transcribed, priming strategy, enzyme type, volume, temperature, and duration of the reverse transcription step.

- Assay performance: PCR efficiency, linear dynamic range, lower limit of detection (LOD), and precision have to be optimized.

- Data Analysis: includes an examination of the raw data, an evaluation of their quality and reliability, and the generation of reportable results.

We performed the experiments starting with RNA obtained from 48 fresh-frozen duodenal biopsies and 49 peripheral blood monocyte samples. Total RNA was reverse-transcribed into cDNA and, after retro-transcription, we carried out a linear pre-amplification step to enhance the low amount of RNA recovered from monocytes. Experiments were performed using the TaqMan® Gene Expression Assay, and the relative expression was calculated with the comparative Ct method.

Finally, we analyzed gene expression using discriminant analysis, which is performed to estimate the contribution of the expression of each gene to distinguish CD patients from healthy individuals and disease controls. The aim of this analysis is to weigh the discriminating capacity of each single gene to obtain a new composite variable, the discriminant score (D-score), which provides a group-specific score for each individual. Wilks' lambda is an estimate of the discriminant capacity ranging from 1 (complete overlap) to 0 (maximum distance). The variable that minimizes the overall Wilks' lambda is entered at each step. According to this analysis, only a few specific genes were selected for discriminating capacity, giving a significant contribution to the Variance ratio F, with a first degree error always less than 0.001.

By multiplying the canonical unstandardized coefficients produced by the analysis to the actual values of the RQ of the candidate genes, a D-score was obtained for each individual. The discriminant score provides a probability of membership to the cases or to the controls groups for each individual. The highest membership probability for each case allows the classification into the diagnostic groups.

2   Materials

2.1   For All the Standard Working Procedures to Be Taken, See Note 1

1. Dynabeads® My Pure™ Monocyte kit (Life Technologies, Foster City, CA), superparamagnetic polystyrene beads coated with a monoclonal human anti-mouse IgG antibody.
2. Ambion® RiboPure™ kit (Life Technologies, Foster City, CA).
3. TRIZOL Reagent.
4. Bromochloropropane (BCP).
5. Glass fiber filter.
6. Low salt buffer.

2.2   RNA Quantification and Quality Control

1. Nanodrop® spectrophotometer.
2. Agarose.
3. 1× Tris/Borate/EDTA buffer (TBE).
4. High Capacity cDNA Reverse Transcription kit (Life Technologies, Foster City, CA).
5. Deoxynucleotide (dNTP) Solution Mix (100 mM).
6. 7900HT Fast Real-Time PCR system.
7. TaqMan® Gene Expression Assay (listed in Table 1).
8. TaqMan® Gene Expression Master Mix (Life Technologies, Foster City, CA).
9. cDNA.
10. SPSS (SPSS Inc., Chicago, IL, USA) and GraphPad Prism 5.0 (GraphPad software, San Diego, CA, USA) software packages.

3   Methods

3.1   Monocyte Isolation

3.1.1   Dynabead Washing

1. Resuspend the Dynabeads in the vial to a homogenous suspension.
2. Transfer the desired volume of Dynabeads to a tube.
3. Add the same volume of Buffer 1 (PBS w/0.1 % BSA and 2 mM EDTA, pH 7.4), or at least 1 ml, and mix.

Table 1
List of taqman assays used for the gene expression analysis

Assay ID	Gene symbol	Gene name	Chromosome	NCBI reference sequence	Target exons	Amplicon length	Position
Hs99999908_m1	GUSB	Glucuronidase, beta	7	NM_000181.3	11–12	81 bp	1913
Hs00611823_m1	TAGAP	T-cell activation RhoGTPase activating protein	6	NM_054114.3	9–10	64 bp	1230
Hs00193878_m1	SH2B3	SH2B adaptor protein 3	12	NM_005475.2	2–3	81 bp	1089
Hs00175260_m1	RGS1	Regulator of G-protein signaling 1	1	NM_002922.3	3–4	115 bp	346
Hs00968436_m1	REL	v-relreticuloendotheliosis viral oncogene homolog (avian)	2	NM_002908.2	10–11	86 bp	1312
Hs0234713_m1	TNFAIP3	Tumor necrosis factor, alpha-induced protein 3	6	NM_006290.2	2–3	63 bp	363
Hs00998604_m1	TNFRSF14	Tumor necrosis factor receptor superfamily, member 14	1	NM_003820.2	6–7	102 bp	987
Hs0222327_m1	IL-21	Interleukin 21	4	NM_001207006.2	3–4	84 bp	406
Hs00944352_m1	LPP	LIM domain containing preferred translocation partner in lipoma	3	NM_001167671.1	8–9	85 bp	1593
Hs00361070_m1	KIAA1109	KIAA1109	4	NM_015312.3	43–44	66 bp	7360
Hs00542477_m1	TNFS14	Tumor necrosis factor (ligand) superfamily, member 14	19	NM_003807.3	4–5	69 bp	685
Hs99999150_m1	IL-2	Interleukin 2	4	NM_000586.3	2–3	89 bp	261

4. Place the tube in a magnet for 3 min and discard the supernatant.

5. Remove the tube from the magnet and resuspend the washed Dynabeads in the same volume of Buffer 1 as the initial volume of Dynabeads.

3.1.2 Preparation of Mononuclear Cells (MNC) from Blood to Obtain Low Platelet Numbers

1. Dilute 10 ml of blood, collected in a preheparinized syringe with PBS pH 7.4 (to a total volume of 35 ml) at 18–25 °C Room Temperature (RT).

2. Add the diluted blood on top of 15 ml of Lymphoprep.

3. Centrifuge at $160 \times g$ for 20 min at RT. Allow to decelerate without brakes.

4. Remove 20 ml of supernatant to eliminate platelets.

5. Centrifuge at $350 \times g$ for 20 min at RT. Allow to decelerate without brakes.

6. Recover MNC from the plasma/Lymphoprep interface and transfer the cells to a 50 ml tube.

7. Wash the MNCs twice with Buffer 1 by centrifugation, first at $400 \times g$ for 8 min at 2–8 °C and second at $225 \times g$ for 8 min at 2–8 °C.

8. Count the collected cells with a Burker chamber (or other method) and resuspend the MNC at 1×10^8 MNC per ml in Buffer 1.
   For critical Steps for Cell Isolation, *see* **Note 2**.

3.1.3 Isolation of Human Monocytes from MNC

1. Transfer 100 µl (*see* **Note 3**) MNC in Buffer 1 to a tube.

2. Add 20 µl Blocking Reagent. Add 20 µl Antibody Mix.

3. Mix and incubate for 20 min at 2–8 °C.

4. Wash the cells by adding 2 ml Buffer 1.

5. Mix the tube and centrifuge at $300 \times g$ for 8 min at 2–8 °C.

6. Discard the supernatant.

7. Resuspend the cells in 900 µl Buffer 1, precooled to 2–8 °C.

8. Add 100 µl prewashed Depletion MyOne Dynabeads and mix.

9. Incubate for 15 min at 2–8 °C and mix (*see* **Note 4**).

10. Resuspend the bead-bound cells by vigorous pipetting 5 times (*see* **Note 5**).

11. Add 1 ml Buffer 1, precooled to 2–8 °C.

12. Place the tube in the magnet (provided from kit) for 3 min and transfer the supernatant to a new tube.

13. Repeat **step 12**. The supernatant contains human monocytes.

3.2 RNA Extraction   Sample acquisition is crucial to minimize experimental variability in mRNA expression experiments that are deeply influenced by sample collection and processing methods.

Nucleic acid extraction is a critical step; choosing an effective and rapid method for tissue or cell disruption is also crucial. The most effective method is determined by the nature of the tissue, the storage method, and the size of the sample. Extraction efficiency depends on adequate homogenization, the type of sample (e.g., in situ tissue vs. log phase cultured cells), target density, physiological status (e.g., healthy, cancerous, or necrotic), genetic complexity, and the amount of biomass processed. To see more precautions to use before working with RNA, *see* **Note 6**. The described procedure, performed by Ambion® RiboPure™ kit, is designed for 5–100 mg tissue samples, $0.1–20 \times 10^6$ cultured cells, or up to 10 cm² of monolayer culture. For samples smaller than 5 mg or 0.1×10^6 cells, *see* **Note 7**.

3.2.1 Cell Disruption and Initial RNA Purification

1. Weigh frozen tissue, and if necessary, break it into pieces smaller than ~50 mg (keeping tissue completely frozen) and homogenize directly in TRI Reagent (*see* **Note 8**).

2. Homogenize samples in 10–20 volumes TRI Reagent (e.g., 1 ml TRI Reagent per 50–100 mg tissue) using standard homogenization procedures.

3. For monocytes (as for any other cells grown in suspension) pellet cells, remove media, then lyse in 1 ml of TRI Reagent per 5×10^6 cells by repeated pipetting or vortexing.

4. Incubate homogenates from both samples and cell cultures with lysis buffer for 5 min at RT. This incubation allows nucleo-protein complexes to completely dissociate.

3.2.2 RNA Extraction

1. Transfer 1 ml of homogenate to a labeled 1.5 ml microcentrifuge tube.

2. Add 100 μl of 1-bromo-3-chloropropane (BCP) (*see* **Note 9**). Cap tubes tightly and vortex at maximum speed for 15 s.

3. Incubate the mixture at RT for 5 min.

4. Centrifuge at $12,000 \times g$ for 10 min at 4 °C to separate the mixture into a lower, red, organic phase (phenol-BCP phase); an interphase; and a colorless, upper, aqueous phase. RNA remains in the aqueous phase while DNA and proteins are in the interphase and organic phase (*see* **Note 10**).

5. Transfer 400 μl of the aqueous phase (top layer) to a new, labeled 1.5 ml microcentrifuge tube.

3.2.3 Final RNA Purification

1. Add 200 μl of 100 % ethanol to 400 μl of aqueous phase from previous step.

2. Vortex immediately at maximum speed for 5 s to avoid RNA precipitation.

3. For each sample, place a Filter Cartridge in one of the Collection Tubes supplied. Transfer the sample to a Filter Cartridge-Collection Tube assembly and close the lid.

4. Centrifuge the assembly at $12,000 \times g$ for 30 s at RT or until all of the liquid is through the filter. Discard the flow-through and return the Filter Cartridge to the same Collection Tube. The RNA is now bound to the Filter Cartridge.

5. Apply 500 µl of Wash Solution to the Filter Cartridge-Collection Tube assembly, and close the lid. Centrifuge for 30 s at RT or until all the liquid is through the filter. Discard the flow-through and return the Filter Cartridge to the same Collection Tube.

6. Repeat the last three steps for a second wash.

7. Centrifuge for 30 s at RT to remove the residual Wash Solution.

8. Add 100 µl of Elution Buffer to the filter column.

9. Incubate at RT for 2 min. Centrifuge for 30 s to elute the RNA from the filter. The RNA will be in the elute, in the Collection Tube.

3.3 RNA Quantification and Quality Control

There are several quantification procedures in common use that produce different results, including spectrophotometry (NanoDrop; Thermo Scientific), microfluidic analysis (Agilent Technologies' Bioanalyzer, Bio-Rad Laboratories' Experion), capillary gel electrophoresis (Qiagen's QIAxcel), or fluorescent dye detection (Ambion/Applied Biosystems' RiboGreen). We recommend comparison of the data obtained with the different methods [5].

We routinely control the quantity of RNA using the Nanodrop® spectrophotometer. The total RNA isolated with previously described procedures should have an A260/A280 ratio of 1.8–2.1. However, RNA with absorbance ratios outside of this range may still function well for quantitative-Reverse PCR (qRT-PCR) or other amplification-based downstream applications. RNA quality may also be analyzed by Agarose gel electrophoresis in Tris/Borate/EDTA buffer (TBE). After the spectrophotometric quantification, we routinely control 100 ng of each RNA sample on 1 % agarose gel electrophoresis, with a reference control size Marker.

3.4 Reverse Transcription

2 µg of total RNA extract (as previously described) from each biopsy and 100 ng RNA from monocytes were reverse-transcribed into cDNA with the High Capacity cDNA Reverse Transcription kit. This kit allows the quantitative conversion of 0.1 to 10 µg of total RNA to cDNA, with a concentration range between 0.002 and 0.2 µg/µl.

3.4.1 Prepare the 2×
Reverse Transcription
Master Mix: Allow the Kit
Components to Thaw
on Ice, per 20 μl Reaction
add the Following Quantity

1. 10× RT Buffer: 2 μl.
2. 10× RT Random Primers: 0.8 μl.
3. 25× dNTP Mix (100 mM): 2 μl.
4. MultiScribe™ Reverse Transcriptase 50 U/μl: 1 μl
5. Nuclease-Free Water: 4.2 μl.

3.4.2 Preparing
the cDNA Reverse
Transcription Reaction

Pipette 10 μl of 2× RT master mix into each well of a 96-well reaction plate or individual tube. Pipette 10 μl of RNA sample into each well, pipetting up and down twice to mix. Seal the plates or tubes and load the thermal cycler (*see* **Note 11**).

3.4.3 Performing
Reverse Transcription:
Program the Thermal
Cycler Conditions

1. Step 1: 25 °C×10 min
2. Step 2: 37 °C×120 min
3. Step 3: 85 °C×5 min
4. Step 4: 25 °C×∞
5. Set the reaction volume to 20 μl, load the reactions into the thermal cycler and start the reverse transcription run.

3.5 qPCR Using
TaqMan Assay

Experiments are performed on the 7900HT Fast Real-Time PCR system using the TaqMan® Gene Expression Assay, and approximately 40 ng of cDNA as described in the protocol.

1. For each sample (to be run in triplicate), pipette the following reagents into a nuclease-free 1.5-ml microcentrifuge tube:
   - 20× TaqMan® Gene Expression Assay: 1 μl
   - 2× TaqMan® Gene Expression Master Mix: 10 μl
   - cDNA template (1–100 ng): 4 μl
   - Nuclease-Free Water: 5 μl

2. Cap the tube and invert it several times to mix the reaction components. Centrifuge the tube briefly. Transfer 20 μl of PCR reaction mix into each well of a 48-, 96-, or 384-well reaction plate. Seal the plate with the appropriate cover. Centrifuge the plate briefly. Load the plate into the instrument.

3. The SDS software (Life Technologies, version 1.4 or 2.4) is used to analyze the raw data and then additional statistical analysis is performed on GraphPad Prism 5.01®.

4. The relative expression is calculated using the comparative ΔCt method. The ΔΔCt method is one of the most popular means of determining differences in concentrations between samples and is based on normalization with a single reference gene. The difference in Ct values (ΔCt) between the target gene and the reference gene is calculated, and the ΔCts of the different samples are compared directly. The expression of each gene is normalized to an endogenous housekeeping gene. The ideal endogenous control should have a constant RNA transcription

level under different experimental conditions and be sufficiently abundant across different tissues and cell types. Although any gene that is stably expressed under the defined experimental conditions can serve as a normalization gene, the selection is most commonly made from constitutively expressed mRNA housekeeping genes, or ribosomal RNAs such as 18S rRNA. Genes such as HPRT [6], GUS, and B2M have shown relative stability across a number of tissues. Thus, there is no universal control gene and it is important to identify the most appropriate endogenous control for a particular cell type and experimental condition. GUSb was chosen as reference gene after it had been determined as the most stable reference gene out of 5 candidates (β-actin, B2M, GAPDH, GUSb, and HPRT1). All gene expression experiments were conducted according to MIQE guidelines [7].

3.6 Statistical Analysis

Using SPSS software, a discriminant multivariate analysis is performed on biopsy gene expression RQ values. Using this approach 5 genes (TNFAIP3, IL-21, c-REL, RGS1, and LPP) were selected for discriminating capacity. The multivariate equation is capable of discriminating celiacs from controls; in fact 92.9 % of individuals were correctly classified efficiently (95 % of controls and 90.9 % of celiacs).

By the same multivariate approach, the expression of 4 candidate genes from monocytes was selected, with a pattern quite similar to that observed in the duodenal tissue. *LPP, c-REL, KIAA1109, and TNFAIP3* genes help to discriminate cases from controls; indeed 91 % of controls and all CD patients were correctly classified.

Finally, we obtained four clustered D-scores, one for each group (Controls, CD, Crohn, and CD on GFD) with no overlap with the active celiacs. The D-Score of active celiac patients was negative in all cases, while it was positive for all the other groups on differentiated clusters. This score produces a group membership probability for each individual, allowing us to correctly classify all controls and CD patients; none of the controls, neither CD on GFD nor Crohn patients were misclassified as CD patients.

Our discriminant function is proposed in an attempt to improve the diagnosis of CD and as a support to limit invasive techniques. Esophago-gastro-duodenoscopy is still the gold standard for the diagnosis of CD, but it can decrease the patient's compliance and is indeed a major bottleneck in developing countries: a simple blood sample, which can be easily dispatched, may help to disseminate the diagnostic coverage to the majority of patients that cannot reach a specialized reference center [7]. In the near future, because of the new ESPGHAN protocol [8], we may have no information about the status of the traditional target tissue in many patients: gene expression on a blood sample may well add safety and sensitivity to a biopsy-free diagnostic protocol, thereby providing a good proxy of the mucosal status.

4   Notes

1. Prepare all solutions with ultrapure water and analytical grade reagents. Prepare and store all reagents at RT. Diligently follow all waste disposal regulations when disposing waste materials.

2. Use a mixer that provides tilting and rotation of the tubes to ensure that Dynabeads do not settle at the bottom of the tube. When incubating Dynabeads and cells, the incubation temperature must be 2–8 °C to reduce phagocytic activity and other metabolic processes. If the temperature is above 2–8 °C, the monocytes will engulf the beads and be depleted from the sample, giving a low recovery of monocytes. Never use less than 100 µl Dynabeads per 1×10^7 MNC sample. It is critical to follow the magnet recommendations to ensure a successful isolation.

3. The protocol described in Subheading 3.1 is based on 1×10^7 MNC; it can be scaled up from $1 \times 10^7 – 5 \times 10^8$ cells.

4. Mix well tube with gentle tilting and rotation several times.

5. Use a pipette with a narrow tip opening (e.g., a 1000 µl pipette tip or a 5 ml serological pipette).

6. Before working with RNA, it is always a good idea to clean the lab bench and pipettors with an RNase decontamination solution (e.g., Ambion RNase*Zap*® Solution). Wear laboratory gloves at all times during this procedure and change them frequently. Gloves protect you from the reagents, and they protect the RNA from nucleases that are present on skin. Use RNase-free pipette tips to handle Wash Solution and Elution Buffer, and avoid putting used tips into the kit reagents.

7. The procedure is compatible with tissues that have been stored in Ambion RNAlater® Solution. Total RNA yield is typically 100–500 µg per 100 mg of tissue, depending on the type of tissue.

8. Larger pieces of tissue, very hard or fibrous tissues, and tissues with a high RNase content must typically be ground to a powder in liquid nitrogen for maximum RNA yield.

9. Alternatively, use 200 µl of chloroform (without isoamyl alcohol) in place of BCP.

10. The volume of the aqueous phase is typically about 60 % of the volume of TRI Reagent used for homogenization.

11. Briefly centrifuge the plate or tubes to spin down the contents and to eliminate any air bubbles. Place the plate or tubes on ice until you are ready to load the thermal cycler.

References

1. Galatola M, Izzo V, Cielo D et al (2013) Gene expression profile of peripheral blood monocytes: a step towards the molecular diagnosis of celiac disease? PLoS One 8(9):e74747

2. Heid CA, Stevens J, Livak KJ, Williams PM et al (1996) Real time quantitative PCR. Genome Res 6:986–994

3. Suzuki T, Higgins PJ, Crawford DR (2000) Control selection for RNA quantitation. Biotechniques 29:332–337

4. Bustin SA (2000) Absolute quantification of mRNA using real-time reverse transcription polymerase chain reaction assays. J Mol Endocrinol 25:169–193

5. http://www.gene-quantification.de/miqe-bustin-et-al-clin-chem-2009.pdf

6. Bustin SA (2005) Real-time, fluorescence-based quantitative PCR: a snapshot of current procedures and preferences. Expert Rev Mol Diagn 5:493–498

7. Greco L, Timpone L, Abkari A et al (2011) Burden of celiac disease in the Mediterranean area. World J Gastroenterol 17:4971–4978

8. ESPGHAN Working Group on Coeliac Disease Diagnosis, ESPGHAN Gastroenterology Committee, European Society for Pediatric Gastroenterology, Hepatology and Nutrition (2012) Guidelines for the diagnosis of coeliac disease. J Pediatr Gastroenterol Nutr 54:136–160

Chapter 12

Cloning Gene Variants and Reporter Assays

Ben Molloy and Ross McManus

Abstract

Recent advances have identified new genetic markers associated with the inheritance of celiac disease. These non-HLA target regions remain to be fully categorized. Investigation of associated SNPs indicates that the causal variants may alter specific gene expression. Thus, closer examination of potential causal variants found within regulatory regions could provide data relating to the mechanistic association. Molecular cloning is an established fundamental tool that enables investigators to examine the differential potential at a variant site. In conjunction with reporter gene assays, SNPs affecting gene expression can be uncovered and contribute to our understanding of the underlying pathogenic mechanisms. This chapter outlines the protocols necessary to clone risk variants and transfect these constructs into a T cell line for reporter assay analysis.

Key words Cloning, SNP, Luciferase, T cell lines

1 Introduction

In the early twentieth century it was widely believed that proteins carried genetic information by virtue of their perceived greater complexity compared to DNA. Gradually this view changed as mounting evidence suggested that DNA carried the genetic blueprint of the organism. In 1944, Avery, McLeod, and McCarthy showed that DNA is the substance responsible for bacterial transformation [1, 2]. In 1952, Alfred Hershey and Martha Chase bolstered this claim with elegant work on bacteriophage clearly showing that DNA, not protein, is responsible for genetic inheritance [3]. Establishing the carrier of genetic information was a crucial step towards genetic manipulation.

In 1973 the collaborative efforts between Stanford University and UCSF produced the seminal work in molecular cloning. Through their research they concluded that their discovery was "potentially useful for insertion of specific sequences from prokaryotic or eukaryotic chromosomes or extra-chromosomal DNA into independently replicating bacterial plasmids" and in doing so,

Anthony W. Ryan (ed.), *Celiac Disease: Methods and Protocols*, Methods in Molecular Biology, vol. 1326,
DOI 10.1007/978-1-4939-2839-2_12, © Springer Science+Business Media New York 2015

established one of the pillars of molecular biology [4]. Molecular cloning harnesses two bacterial properties: restriction endonucleases and independently replicating plasmids. These tools can be used to create recombinant vectors carrying manipulated genetic code. These newly created DNA vehicles can be introduced into bacterial hosts allowing them to be replicated and propagated with high fidelity indefinitely.

Cloning technology has developed enormously since then, with myriad applications in molecular biology and the life sciences. The focus of this chapter is on the analysis of variation at putative transcription factor binding motifs. A powerful tool to investigate the potential for DNA sequences to influence gene expression has been the development of reporter vectors such as those equipped with firefly luciferase, which fluoresce in proportion to the level at which the luciferase is transcribed. Luciferase assays are rapid, inexpensive, very sensitive and utilize readily available nonradioactive substrates. In addition, luciferin (luciferase's substrate) can diffuse across mammalian cytoplasmic membranes allowing for the detection of luciferase activity in intact cells [5].

Celiac disease (CD) is a chronic small intestine immune-mediated enteropathy induced by dietary gluten in genetically predisposed individuals [6]. Recent research has focused on identifying non-HLA genomic regions and deciphering the contribution they make to the inheritance and development of CD [7]. While causal variants have not been conclusively identified, 53 % of CD-associated SNPs are genetic variants for which different genotypes correlate with differences in expression levels in at least one physically close gene. The number of these variants is significantly greater than would be expected by chance alone [8]. This implies that many of these risk SNPs may contribute to the development of CD through altered gene expression. Gene expression can be affected by subtle differences in promoter and enhancer regions, and 5′ and 3′ UTRs that may harbor transcription factor motifs, among other mechanisms. Thus variants found to lie within these regions could potentially contribute to disease susceptibility. Cloning risk SNPs found within promoters or UTRs in expression reporter constructs is one means of determining whether or not the variants affect gene expression levels. These data could aid in refining the search for causal SNPs and consequently identify non-HLA genes involved in the pathogenesis of CD or similar diseases.

This chapter will outline the protocols necessary to clone risk variants and transfect these constructs into a T cell line. As CD is a T cell-driven ailment, using a T cell line such as Jurkat, HUT-78 or MOLT-4 allows assays to be performed in biologically relevant cellular environment. However consideration should be given to the use of other cell lines such as B cells or the widely used HEK293.

2   Materials

Prepare and store all reagents at room temperature unless otherwise stated.

2.1   Cloning of Site into TOPO TA Cloning Vector

1. Taqman Universal Mix (Applied Biosystems).
2. Custom SNP Assay (Applied Biosystems).
3. Adhesive Plate Covers (Applied Biosystems).
4. Custom Primers (IDT).
5. Taq Polymerase (Qiagen).
6. 10× Buffer (Qiagen).
7. dNTPs (Qiagen).
8. Sterile, DNase-, RNase-free PCR Tubes (Sarstedt).
9. Thermal Cycler.
10. Agarose Powder (Sigma Aldrich).
11. 10× TBE (Sigma Aldrich).
12. Ethidium Bromide Solution (Sigma Aldrich).
13. Conical Flask.
14. Microwave.
15. Owl™EasyCast™ Gel Electrophoresis System (Thermo Scientific).
16. Loading Dye (Qiagen).
17. 100 bp DNA Ladder (Qiagen).
18. 1 kb DNA Ladder (Qiagen).
19. Geldoc (Bio-Rad).
20. LB Agar Capsules (MP Biomedicals).
21. LB Medium Capsules (MP Biomedicals).
22. Ampicillin 10 mg/ml Solution (Sigma Aldrich).
23. Petri Dishes (Sarstedt).
24. TOPO® TA Cloning® Kit (Invitrogen/Life Technologies).
25. Inoculating Disposable Loops (Fisher Scientific).
26. Disposable Spreaders (Fisher Scientific).
27. X-Gal Substrate (Sigma Aldrich).
28. QIAprep® Spin Miniprep kit (Qiagen).

2.2   Subcloning into pGL3 Luciferase Reporter Vector

1. Restriction Enzymes, *XhoI*, *SacI*, and Buffers (New England Biolabs).
2. Genecatcher disposable gel excision tips, 6.5 mm×1.0 mm (Gel Company).
3. QIAquick® Gel Extraction Kit (Qiagen).

4. T4 DNA ligase, Ligation Buffer (New England Biolabs).

5. pGL3 Promoter Luciferase Reporter Vector (Promega).

6. NEB Turbo Competent *E. coli* (New England Biolabs).

7. SOC medium (New England Biolabs).

8. Water Bath.

9. Incubator with agitator.

10. QIAprep® Spin Midiprep Kit (Qiagen).

2.3 Cell Culture, Transfection, and Luciferase Assays

1. Jurkat Cells (HPA Cultures).

2. RPMI 1640 (Sigma Aldrich).

3. Fetal Bovine Serum (Gibco/Life Technologies).

4. DMSO (Sigma Aldrich).

5. Mr. Frosty Freezing Container (Fisher Scientific).

6. T75 Cell Culture Flasks with ventilated caps (Sarstedt).

7. 2 mm Gap Cuvettes (VWR).

8. BTX Square Electroporator.

9. Sterile 24 well plates (Sarstedt).

10. Luminometer 96 well plates (Promega).

11. Renilla Control Vector (Promega).

12. Dual-Luciferase® Reporter Assay Kit (Promega).

13. Luminometer/Plate Reader (VICTOR2™).

3 Methods

Perform all steps at room temperature unless otherwise indicated.

3.1 Cloning of Site into TOPO® TA Cloning® Vector (See Note 1)

The first step is to find celiac homozygote individuals for each variant of the polymorphism. To assay differences, both variants must be cloned. In cases where the MAF is very low, amplification of heterozygotes is advisable and often necessary.

3.1.1 Genotyping DNA Samples

1. Load 1 μl of each DNA sample to be analyzed in a separate well of a 96 well plate.

2. Add 9 μl of the following solution to each well; 500 μl Taqman Universal Mix, 25 μl of SNP Assay and 375 μl deionized H_2O.

3. Seal plate with adhesive cover.

4. Run on a Real-Time PCR Machine.

5. The data will be presented as an X-Y Plot, where the X and Y axis correspond to the fluorescence for the probe for each variant. Values clustered near the top left and bottom right of the plot are homozygotes. Those values clustered between are heterozygotes.

Table 1
Standard PCR reaction quantities and volumes

PCR reagents	Volumes (µl)
Template DNA (10–100 ng)	1
dNTP (2.5 mM each)	1
10× Buffer	1
ddH$_2$O	4.8
Primer 1 (20 µM)	1
Primer 2 (20 µM)	1
Taq polymerase (*see* **Note 5**)	0.2
Total volume	10

3.1.2 Primer Design

1. Open the free online software program "Primer 3." (There are a number of free primer generating programs online.)
2. Paste desired sequence into the program in FASTA format.
3. Alter the "Product Size Range" as appropriate (*see* **Note 2**).
4. The "Primer GC%" can be altered. As a baseline, 30–70 % range is suitable for most primers (*see* **Note 3**).
5. Click "Pick Primers."
6. Repeat until satisfied with primer choices.
7. Additionally, restriction sites can be added onto the primer 5′ ends to facilitate direct cloning into a reporter vector (*see* **Note 4**).

3.1.3 PCR

1. Add the following reagents to a PCR tube according to Table 1.
2. Run sample(s) on a thermal cycler with a predetermined optimal annealing temperature (*see* **Note 6**).

3.1.4 Gel Electrophoresis

The following is the protocol for making a 1 % agarose gel for examining PCR products and digested plasmids.

1. Add 1.5 g of agarose powder to 150 ml of 1× TBE (TAE may also be used) in a conical flask.
2. Heat the solution in a microwave for approximately 90 s. Stop every 20 s to mix and to prevent from boiling over.
3. Allow the solution to cool to 55 °C, then add 12 µl of Ethidium Bromide Solution (10 mg/ml). Handle Ethidium Bromide with care.
4. Swirl the flask to assimilate the dye while avoiding generating bubbles, then pour carefully but purposefully into a gel tray with a well comb in place.

5. Leave for 40 min to solidify.

6. Immerse the gel in 1× TBE.

7. Prepare samples by adding 1 μl of loading dye for every 5 μl of sample (*see* **Note 7**).

8. Load 6 μl of an appropriate ladder into the gel and add 12 μl of each sample.

9. Attach the leads to a power source and run at 77 V for 2 h or 67 V for 3–4 h (*see* **Note 8**).

10. Image the gel under UV light.

3.1.5 LB Agar Plates

1. Add four LB agar medium capsules to 100 ml of ddH$_2$O (*see* **Note 9**).

2. Shake vigorously until well dispersed.

3. Autoclave the solution.

4. Allow it to cool to approximately 55 °C before pouring into Petri dishes. Ampicillin or another antibiotic may be added prior to pouring (50 μg/ml). Disperse added agents by swirling the contents; avoid generating bubbles in the solution.

5. Once solidified and dry, turn upside down and store at 4 °C. Suitable for culturing up to 1 week.

3.1.6 LB Medium

1. Add ten LB medium capsules to 400 ml of ddH$_2$O.

2. Shake vigorously until well assimilated.

3. Autoclave the solution.

4. Store at room temperature.

5. Add antibiotic to aliquots only immediately prior to usage (50 μg/ml).

3.1.7 Ligation and Transformation

1. To a PCR tube add 1 μl of PCR product, 1 μl of salt solution, 1 μl of TOPO® vector, and 3 μl of ddH$_2$O (*see* **Note 10**).

2. Mix the reaction gently and rest for 5 min.

3. After 5 min place on ice and proceed to Transformation.

4. Add 2 μl of the cloning reaction to a vial of *E. coli* One Shot® competent cells and mix gently and quickly to avoid letting the suspension heat up. Do not pipette up and down.

5. Incubate on ice for 30 min.

6. Heat shock the cells for 30 s at 42 °C.

7. Immediately transfer vials to ice and allow to rest for 5 min.

8. Add 250 μl of S.O.C medium at room temperature to the vial(s), lay it horizontally, and allow it to shake vigorously (250 rpm) for 1 h at 37 °C.

9. At this time, spread 40 μl of 40 mg/ml X-gal solution on each LB plate and incubate at 37 °C until ready for use (*see* **Note 11**).

10. Pipette 50 µl of the transformation(s) on prewarmed LB plates with ampicillin and X-gal substrate.

11. Incubate plate (upside down) at 37 °C overnight.

12. White colonies on the plate have the vector with the insert.

3.1.8 Vector Extraction

1. Pick a white colony and inoculate in 10 ml of LB medium with ampicillin (5 µg/ml).

2. Incubate at 37 °C with vigorous shaking overnight.

3. Centrifuge the culture at 8000 rpm (*see* **Note 12**) for 3 min to harvest bacterial pellet.

4. Follow the instructions as detailed in the QIAprep® Spin Miniprep Handbook for isolation of plasmid DNA.

5. Elute DNA in 50 µl of elution buffer (*see* **Note 13**) and quantify using a Nanodrop or similar device.

6. Sequence the TOPO® vector (*see* **Note 14**).

3.2 Subcloning into PGL3 Luciferase Reporter Vector

The next stage is to remove the insert from the TOPO® vector and ligate it into the reporter vector, pGL3 PRO. When cutting both plasmids, the same restriction endonucleases must be utilized to benefit from complementarity. While one enzyme can be used, this protocol will detail a double digest using two different restriction enzymes. This method has two benefits over a single digest. First, the cut vector cannot religate assuming the enzymes don't create cohesive ends and, second, the orientation of the insert can be controlled.

3.2.1 Restriction Digestion of Plasmids

This protocol will detail the reagents necessary for a double digest on TOPO® constructs and pGL3-Promoter luciferase reporter vector. As discussed, double digests have benefits over single digests. Studying the multiple cloning sites (MCSs) of these vectors provides a number of choices. I have chosen *XhoI* and *SacI*.

1. For each digest add the following: 20 µl(~1 µg) of DNA (vector), 5 µl of NEB Buffer 1, 1 µl of *SacI*, 1 µl of *XhoI*, 1 µl of BSA, and 22 µl of ddH$_2$O (*see* **Note 15**).

2. Incubate at 37 °C from 1 h to overnight (*see* **Note 16**).

3.2.2 Gel Electrophoresis and Purification of Bands

1. The protocol for making and running the gel is detailed in Subheading 3.1.3. The only amendment is to use a larger well comb so that a larger volume can be loaded.

2. The gel is exposed under UV light to reveal the inserts from the TOPO® constructs and cut pGL3 vector.

3. Using an excision tip or sharp blade, the inserts and cut reporter vector are removed from the gel.

4. The gel pieces are weighed and recorded.

5. Follow the instructions in the QIAquick® Gel Extraction Kit Protocol (using a microcentrifuge).

6. Elute in the smallest volume possible, 30 μl (*see* **Note 17**).

3.2.3 Ligation and Transformation

The next step involves ligating the excised insert from the TOPO® constructs into the pGL3 Luciferase reporter vector. This ligation reaction makes use of complementary sticky ends generated by restriction endonuclease digestion and is catalyzed by a ligase enzyme such as T4 DNA ligase rather than topoisomerases as with the TOPO vector. This reaction requires an excess of insert to vector and the exact amount is dependent on insert and vector size. Using a suggested ratio of 10:1 (*see* **Note 18**) the formulae for calculating the amount of insert is given; Mass of Insert = 10*(insert length/vector length)*vector mass. Once these calculations have been made, follow the steps below to perform a ligation reaction.

1. Add the following to a PCR tube: 1 μl Ligation Buffer (10×), 1 μl T4 DNA Ligase, 1 μl pGL3 vector, Xμl Insert, Yμl ddH$_2$O to a final volume of 10 μl (*see* **Note 19**).

2. Incubate at room temperature for at least 3 h (*see* **Note 20**).

3. Incubate at 65 °C for 5 min to inactivate the T4 DNA ligase.

4. Proceed directly to transformation or store samples at 4 °C until ready for use.

5. New England Biolabs (NEB) Turbo® Competent *E. coli* (or other suitably competent host cells) are thawed on ice for 10 min.

6. After 10 min or when the last bit of ice has dissolved in the vial(s), add half the volume of the ligation reaction to the vial(s) (*see* **Note 21**).

7. A positive control (pUC19) and a negative control (containing just competent cells) should also be used.

8. Flick and rotate the vial(s) *gently* to mix and incubate on ice for 30 min.

9. Heat shock at 42 °C for 30 s and place back on ice for 5 min (*see* **Note 22**).

10. Add 250 μl of NEB SOC medium to each vial and incubate at 37 °C for 1 h with vigorous shaking.

11. Spread 100 μl from each vial on a selective agar plate containing ampicillin (5 μg/ml) and incubate overnight at 37 °C (*see* **Note 23**).

3.2.4 Colony Selection and Screening

Unlike TOPO® cloning, pGL3 has no marker to determine whether the insert is present or not. The only thing we can say of the colonies on the plate is that they have resistance to ampicillin and thus carry the resistance gene on pGL3. However, the double digest performed should ensure that any colonies present have the reporter vector with the desired insert; since the enzymes used to cut the vector do not produce compatible cohesive ends, the vector should not be able to religate on itself. If the orientation of the insert in the TOPO® construct was confirmed, then the insert should be in the same orientation. If it wasn't confirmed, then the insert may not be in the desired orientation. First one must prove that the insert is intact and present. This can be achieved through a simple restriction digest using the same enzymes from the double digest. Sanger sequencing is the most convenient method for proving the orientation.

1. Pick a single, independent white colony.

2. Streak it on a fresh ampicillin agar plate and incubate at 37 °C overnight.

3. Extract the vector construct from the bacteria following protocols already detailed.

4. Perform a restriction digest and run the product on a 1 % agarose gel.

5. Send a sample for DNA sequencing (*see* **Note 24**).

3.3 Transfection into a Mammalian Cell Line

At this point, the desired sequence with each variant has been cloned successfully into a luciferase reporter vector and is ready to be tested in a cellular environment. If you are testing the ability of a transcription factor to recognize a site in your construct, then your cell line must express that protein at either under basal conditions or post activation. If gene expression cannot be induced, then a second construct containing this gene will need to be co-transfected. For the purpose of this protocol, we will assume that your gene of interest (i.e., transcription factor) is expressed at a high basal level in your cell line.

3.3.1 Cell Culture

Jurkat cells are cultured at 37 °C and 5 % CO_2 in a humidified atmosphere in T75 culture flasks. The cells grow in suspension in RPMI with 10 % Fetal Bovine Serum (FBS). The cells are not difficult to maintain but do tend to clump and grow aggressively. This means that cultures should be split two to three times per week. Prior to performing experiments, cells should be separated by pipetting up and down several times. The protocol below is for starting, splitting, and freezing Jurkat cultures.

1. Take the frozen vial containing Jurkat cells and thaw rapidly using a water bath at 37 °C.

2. Add the total volume to 10 ml of RPMI with 10 % FBS and incubate at 37 °C.

3. When splitting, centrifuge the suspension at 1500 rpm for 5 min.

4. Resuspend pellet in 2 ml of media and add 1 ml to one flask containing 9 ml of media and the second 1 ml to a second identical flask. Incubate at 37 °C.

5. When freezing cells, aliquot 1×10^6 cells in 1 ml of RPMI with 20 % FBS, 10 % DMSO.

6. Place a 1 ml vial in a Mr. Frosty (Isopropanol bath) and put at −80 °C and leave overnight (*see* **Note 25**).

7. Remove the vial and place in normal storage conditions at −80 °C or for longer storage, put in liquid nitrogen.

3.3.2 Transfection

Primary T cells and T cell lines are difficult to transfect using lipofectamine and other reagents that interfere with cell membrane integrity. Electroporation is the best method for transfecting T cells with constructs and siRNAs [9, 10].

1. Count Jurkat cells and resuspend in RPMI with 20 % FBS.

2. Make 100 μl aliquots each containing 1.5×10^6 cells.

3. Add 2 μg of experimental vector and 50 ng of Renilla control vector (40:1 ratio) (*see* **Note 26**). In addition, an empty pGL3 vector should be used as a baseline control.

4. Add the cell-construct mixture to a 2 mm Gap cuvette and electroporate at 150 V for 12 ms.

5. Incubate overnight.

6. Stimulate with PMA for 4 h (*see* **Note 27**) and lyse cells in Promega 1× Passive Lysis Buffer (PLB).

3.3.3 Luciferase Assay

1. Follow detailed instruction in Promega's Dual-Luciferase® Reporter Assay System protocol.

2. Measure luciferase and renilla values on a luminometer plate reader.

3. Normalize luciferase values against renilla values for each sample in triplicate. Then normalize these values against the empty vector to obtain an accurate measure of luciferase activity.

4. Repeat the experiment on three occasions to minimize effects of stochastic variation.

4   Notes

1. PCR products can be cloned in two ways, directly into reporter vectors or subcloned using an intermediary vector. There are advantages and disadvantages with both approaches. Direct cloning allows one to add desired restriction sites to the amplification primers and it is a more rapid form of cloning. Subcloning takes longer and you are restricted in the number of restriction sites but it will work reliably, sticky ends will be created efficiently and symmetrically, and sequencing can be performed to ascertain orientation of the insert at an early stage.

2. It is advisable to keep insert sizes below 1 kb. The TOPO®
   TA Cloning® system is not efficient for insert sizes greater than
   1 kb.

3. The GC% of a primer is directly related to its annealing temperature to template DNA during PCR. The higher the GC%
   content, the higher the melting temperature.

4. Taq DNA polymerase must be used as it creates a nontemplate
   overhanging A residue which is exploited by the TOPO kit for
   ligation of PCR products.

5. If you wish to add restriction sites to facilitate direct cloning
   into a reporter vector, then it is advisable to add an additional
   6–10 bases after the restriction site. This will protect the site
   and provide extra stability and is a requirement of some
   restriction enzymes to allow proper binding and restriction of
   target sites.

6. Prior to performing PCR for cloning, a temperature gradient
   PCR should be performed to determine the optimal annealing
   temperature which is principally determined by the GC% content of the primers. Choose 12 evenly spaced temperatures
   that range between 55 °C and 65 °C. With a GC% content of
   30–70 %, the annealing temperature should fall within the
   specified temperature range.

7. When using a small well comb, the maximum amount of sample loaded should not exceed 12 μl.

8. Sometimes it is necessary to run a gel at a lower voltage for a
   longer time if gel bands are appearing streaked or running into
   each other. This tends to be an issue with smaller fragments.
   However running the gel at 60 V for 2 h under supervision
   should be adequate in most scenarios.

9. This quantity is enough for approximately 4 plates. Multiply
   quantities to suit individual needs.

10. The TOPO® Kit supplies a salt solution to be used for ligation.
    The manufacturers state that it increases the number of trans-
    formants two- to threefold. They also recommend leaving the
    cloning reaction rest for at least 5 min, but not more than
    10 min at room temperature prior to placing on ice. The vol-
    ume of PCR product is also variable. It can range from 0.5 to
    4 μl. Ensure to adjust ddH$_2$O volume accordingly.

11. This step can be performed earlier as sometimes it can take
    longer than 1 h for the X-gal solution to be absorbed into
    the LB agar if the plates have not been dried sufficiently. The
    TOPO® Cloning Vector has a LacZ gene which is disrupted
    when an insert is present and is intact when an insert is absent.
    When disrupted the bacteria can't produce beta-galactosidase
    (LacZ) and as such can't digest the artificial substrate. The

colonies appear white. Just because a bacterium has taken up the vector doesn't mean it has the insert; although if appropriate measures (such as the use of restriction enzymes that give rise to non-compatible ends, or dephosphorylation of the cut vector) are used, the numbers of vectors without inserts can be minimized. Where the vector is capable of religating on itself, antibiotic resistance isn't sufficient for choosing insert-positive colonies.

12. Most centrifuges that take 50 ml falcons cannot attain these speeds. I have found that centrifuging at 5000 rpm for 10 min pellets the bacterial cells effectively. It is also advisable prior to centrifugation to take a 500 µl aliquot. This can be frozen at −80 °C with a 1:1 ratio of glycerol. Making stocks saves time if more plasmid is ever required.

13. Eluting in a small volume allows for a concentrated solution. It is much easier to dilute a solution than to concentrate it.

14. Sanger sequencing is both cost effective and efficient. It can be outsourced for a modest price. Restriction enzymes can be used to confirm predicted insert size. However, sequencing is more reliable and faster and can identify any unintended mutations introduced during PCR amplification which is a relatively low fidelity process.

15. Double digests must contain a buffer that both enzymes can work efficiently in. Check supplier recommendations.

16. It is possible to leave the restriction digest overnight, although technically they can be left for just 1 h, depending on the amount of DNA and enzyme present. Check supplier technical information.

17. Extracting DNA fragments from gels has its advantages and disadvantages. The main advantage is that the product returned should be pure, uniform, and free of damaging salts that could perturb a ligation reaction. The main disadvantage is that yields from gel purification tend to be low. For this reason, loading as much digested vector as possible into each well and/or aggregating wells is crucial. Expect yields to be as low as 30 % of the originally loaded amount.

18. This ratio may need to be optimized. A 10:1 ratio is a good starting point, in certain cases an even higher ratio may be required.

19. The amount of insert ("X"µl) will have been determined by the calculation detailed in the brief introduction. The amount of ddH$_2$O ("Y"µl) is calculated by the amount required to achieve a final volume of 10 µl.

20. New England Biolabs retail a "Quick Ligation" kit which carries out the ligation at 25 °C in 5 min. However as the ligation buffer contains Peg-6000, it is not suitable for later transfections employing electroporation.

21. Usually an excess of cells is supplied in each NEB vial. The cells can be split between two sterile 1 ml tubes if preferred. Ensure that the samples are always kept on ice and are not allowed to heat up.

22. Flick the vials gently and it cannot be stressed enough how important it is to follow the protocol exactly. The times and temperatures given, particularly for the heat shock stage, must be exact.

23. Streak plates beside a Bunsen burner. It is the most convenient way to maintain an aseptic environment while the plates are exposed to the air.

24. While Sanger sequencing is the preferred method has its own limitations. The maximum read for a single reaction is 800 bases, but in reality it is closer to 600. The plasmid can be sequenced in either direction using vector associated or designed primers. Thus, determining orientation is normally possible in all cases. In addition, once your construct has been confirmed, freeze some bacterial cells containing your pellet at −80 °C and perform a Midiprep (Qiagen) to obtain a larger amount of construct. A midiprep is essentially a large-scale miniprep and the protocol is freely available from Qiagen.

25. As a general rule with cell lines, they should be frozen slowly and thawed rapidly.

26. This ratio will need to be optimized. It is generally advisable to perform an optimization experiment where ratios of 5:1, 10:1, 25:1, and 40:1 are tested. The control vector is important for normalizing raw luciferase values. However as both vectors have strong promoters, the *trans* effects must be considered [11]. To ensure that these effects are minimized, the amount of control vector should be as low as possible. Promega even suggests ratios greater than 50:1 in appropriate cases.

27. You may not need to stimulate your cell line. Although cell lines are valuable models, their genomes are mutated to a greater or lesser extent and aberrant gene expression is common. Only stimulate with PMA or a similar activation agent if it is required to raise expression of your target gene or is otherwise necessary.

Acknowledgements

The authors acknowledge Science Foundation Ireland (SFI) grant 09/IN.1/B2640 to R.M.M. We would also like to thank Dr. Richard Anney, Dr. Graham Kenny, Dr. Emma Quinn, and Dr. Anthony W. Ryan for their assistance and advice.

References

1. Avery OT, MacLeod CM, McCarthy M (1944) Studies on the chemical nature of the substance inducing transformation of pneumococcal types. J Exp Med 79:137–158

2. Lederberg J (1994) The transformation of genetics by DNA: An anniversary celebration of Avery, Macleod and McCarthy (1944). Genetics 36:423–426

3. Hershey A, Chase M (1952) Independent functions of viral proteins and nucleic acid in growth of bacteriophage. J Gen Physiol 36:39–56

4. Cohen SN, Chang ACY, Boyer HW, Helling RB (1973) Construction of biologically functional bacterial plasmids in vitro. Proc Natl Acad Sci U S A 70:3240–3244

5. de Wet JR, Wood KV, Deluca M, Helinski DR, Subramani S (1987) Luciferase gene: structure and expression in mammalian cells. Mol Cell Biol 7(2):725–737

6. Ludvigsson JF, Leffler DA, Bai JC, Biagi F, Fasano A, Green PH, Hadjivassiliou M, Kaukinen K, Kelly CP, Leonard JN et al (2013) The Oslo definitions for coeliac disease and related terms. Gut 62(1):43–52

7. Dubois PC, Trynka G, Franke L, Hunt KA, Romanos J et al (2010) Multiple common variants for coeliac disease influencing immune gene expression. Nat Genet 42:395–402

8. Abadie V, Sollid LM, Barreiro LB, Jabri B (2011) Integration of genetic and immunological insights into a model of celiac disease pathogenesis. Ann Rev Immunol 29:493–525

9. Hurez V, Hautton RD, Oliver J, Matthews RJ, Weaver CK (2002) Gene delivery into primary T cells: overview and characterization of a transgenic model for efficient adenoviral transduction. Immunol Res 26:131–141

10. June CH, Blazar BR, Riley JL (2009) Engineering lymphocyte subsets: tools, trials and tribulations. Nat Rev Immunol 9:704–716

11. Farr A, Roman A (1992) A pitfall of using a second plasmid to determine transfection efficiency. Nucleic Acids Res 20:920

Chapter 13

Epigenetic Methodologies for the Study of Celiac Disease

Antoinette S. Perry, Anne-Marie Baird, and Steven G. Gray

Abstract

Epigenetic regulation of gene expression is an important event for normal cellular homeostasis. Gene expression may be "switched" on or "turned" off via epigenetic means through adjustments in the architecture of DNA. These structural alterations result from histone posttranslation modifications such as acetylation and methylation on key arginine and lysine residues, or by alterations to DNA methylation. Other known epigenetic mechanisms invoke histone variant exchange or utilize noncoding RNAs (lncRNA/miRNA). Drugs which can target the epigenetic regulatory machinery are currently undergoing clinical trials in a wide variety of autoimmune diseases and cancer.

Here we describe RNA isolation and the subsequent Reverse Transcriptase Polymerase Chain Reaction (RT-PCR) methods, post-epigenetic drug treatment, to identify genes, which may be responsive to such epigenetic targeting agents. In addition, we depict a chromatin immunoprecipitation (ChIP) assay to determine the association between chromatin transcription markers and DNA following pretreatment of cell cultures with a histone deacetylase inhibitor (HDi). This assay allows us to determine whether treatment with an HDi dynamically remodels the promoter region of genes, as judged by the differences in the PCR product between our treated and untreated samples. Finally we describe two commonly used methodologies for analyzing DNA methylation. The first, methylation-sensitive high resolution melt analysis (MS-HRM) is used for methylation screening of regions of interest, to identify potential epigenetic "hotspots." The second, quantitative methylation specific PCR (qMSP) is best applied when these hotspots are known, and offers a high-throughput, highly sensitive means of quantifying methylation at specific CpG dinucleotides.

Key words RNA, RT-PCR, Histone deacetylase, Chromatin immunoprecipitation, DNA methylation

1 Introduction

It was in 1942 that Waddington coined the term epigenetics [1]. It is defined as the study of heritable changes in gene expression without a resultant change in DNA sequence. The term stems from the Greek word "epi" meaning "upon." Thus one envisages another layer of heritable information in addition to the DNA code, in which biochemical modifications sit "upon" chromatin— both DNA and histones, influencing the transcriptional activity of the genome. The conformation of our DNA dictates the

Anthony W. Ryan (ed.), *Celiac Disease: Methods and Protocols*, Methods in Molecular Biology, vol. 1326, DOI 10.1007/978-1-4939-2839-2_13, © Springer Science+Business Media New York 2015

expression patterns of genes, as it determines when and where along the DNA sequence that transcription factors can access the DNA to begin the process of transcription. These conformational changes are both heritable and stable and play an important role in regulating gene expression. Cyr and Domann conceptualized epigenetics when they used the example of baking a cake [2]. By varying the amount of the same inputs (flour, butter, and sugar), you can alter the phenotype of the product "cake or cookie" [2]. Epigenetic mechanisms currently known involve the following: DNA methylation, histone posttranslational modifications (PTMs), histone variant deposition, imprinting, and noncoding RNA (ncRNA) [3–5].

Human DNA is approximately 2 M in length but is packaged into the nucleus, which is approximately 6 μM in diameter. To achieve this amazing feat, the DNA is tightly packaged and condensed (Fig. 1). The DNA binds to and is wrapped around a complex of histones called the nucleosome. Each individual nucleosome core consists of two molecules each of histone H2A, H2B, H3, and H4. This histone octamer forms a protein core around which double-stranded DNA is bound. Other linker histones or proteins then associate with and progressively fold the nucleosome into higher order structures to form chromatin.

Histone proteins are among the most highly evolutionarily conserved proteins. Each histone has an N terminal amino acid "tail" which extends out from the DNA-histone core. These tails are subjected to several modifications that control crucial parts of chromatin structure and function, and the "histone code" is a well-established hypothesis describing the idea that specific patterns of PTMs to histones act like a molecular "code" recognized and used

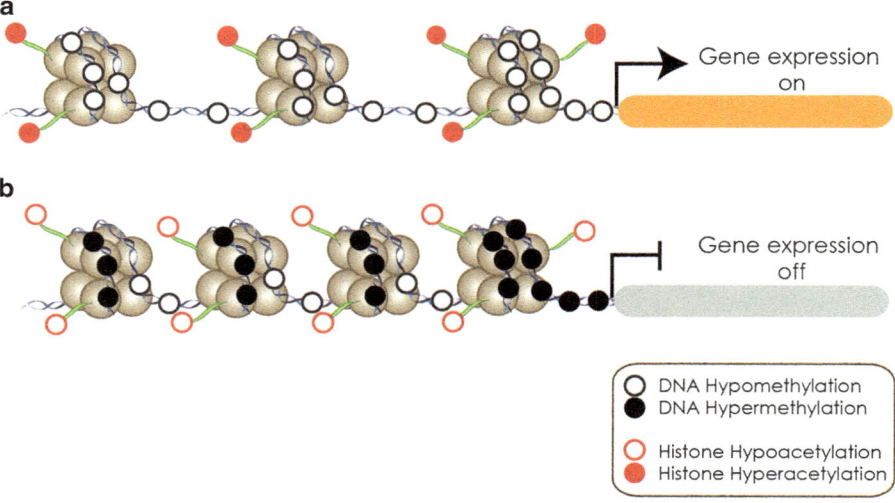

Fig. 1 The role of epigenetic modifications in regulating gene expression

by non-histone proteins to regulate specific chromatin functions. These modifications include acetylation, methylation, phosphorylation, sumoylation, and ubiquitination, and various families of proteins have been identified which function to place or remove these PTMs. The best studied of these families are the Lysine-acetyltransferases/K-acetyltransferases (KATs), histone deacetylases (HDACs), K-methyltransferases (KMTs), and K-demethylases (KDMs) [6].

DNA methylation is probably the best understood of all the epigenetic modifications. The addition of a methyl group from the universal methyl donor *S*-adenosylmethionine to the fifth carbon of cytosine in cytosine–phosphate–guanine (CpG) dinucleotides converts cytosine to 5-methylcytosine (5mC) [7]. DNA methylation is mediated by a family of methyltransferase enzymes, of which three catalytically active members are known (DNMT1, DNMT3A, and DNMT3B). DNMT1 is essential for maintaining patterns of methylation over replicative time by copying the pattern of the parent strand onto the new strand during DNA replication [8]. DNMT3A and DNMT3B possess the capacity to target unmethylated CpG sites and therefore function in establishing de novo methylation patterns [9, 10]. Recent studies of human embryonic stem cells have also shown methylation to exist in certain restricted non-CpG contexts [11].

Most CpG dinucleotides (estimated at between 60 and 90 %) are methylated in the adult vertebrate genome, leading to the spontaneous deamination of 5-methylcytosine to thymine. As a result, our genome overall is underrepresented in CpG. However, CpGs are nonrandomly distributed; approximately 1 % of our DNA consists of short, CpG-rich sequences termed CpG islands [7, 12, 13]. CpG islands are co-located at the promoters of approximately 60 % of all human genes and are typically unmethylated, thus avoiding mutation to TpG [4]. Hypermethylation of such promoter CpG islands is associated with gene inactivation. This occurs through inhibiting transcription initiation, both directly (by excluding transcription factor binding) and indirectly, by attracting methyl-CpG-binding domain proteins that interact with histone deacetylases and chromatin remodelling factors, thus inducing chromatin condensation, rendering the promoter inaccessible for transcription (Fig. 1). Nucleosome positioning is also proposed to orchestrate the landscape of global DNA methylation patterns [4, 14].

Other important regulators of epigenetic gene expression include noncoding RNAs (ncRNAs) such as microRNAs (miRNA) and long noncoding RNA (lncRNA). miRNAs are small, approximately 22 nucleotide ncRNAs that regulate gene expression through posttranscriptional silencing of target genes, by binding to complementary sequences on target messenger RNA transcripts (mRNAs), resulting in either mRNA degradation or translational repression and gene silencing. Their primary roles are to regulate the self-renewal, differentiation, and division of cells and their levels are frequently altered in cancer and autoimmune conditions [15, 16]. Crucially however, miRNAs themselves can be epigenetically regulated, while a specific subset of miRNAs has been shown to directly regulate the epigenetic machinery (leading to the term epi-miRNAs) [17, 18]. LncRNAs are emerging as highly important molecules in orchestrating epigenetic gene regulation, found to modulate chromatin structure, and mediate enhancer-promoter looping among many other activities [19]. In terms of inflammatory bowel disease (IBD), there is evidence in the literature to suggest a link between aberrant epigenetics and its pathogenesis. In Genome Wide Association Study (GWAS) studies, a number of DNA methyltransferase enzymes such as DNMT3A and -3B were found to play a role in Crohn's disease [20, 21]. Evidence of genes aberrantly methylated in IBD conditions includes e-cadherin (*CDH1*) and *TMEFF2* (also known as *HPP1*) among others [22, 23]. An early suggestion that methylation may be important in susceptibility to celiac disease came from a study of HLA genes. It is well established that celiac disease is twice as frequent among female as males, and in an investigation of paternal/maternal inheritance of HLA-DQ disease-predisposing haplotypes, Mazzilli and colleagues observed a major distortion in the DR3-DQ2 transmission from fathers to daughters, suggesting that parent-specific epigenetic modifications in the two genders may be involved [24]. More recently, the DNA methylation status of eight genes associated with the NFkB signalling pathway was examined in celiac intestinal mucosa. Altered DNA methylation was observed in patients with celiac disease even after more than 2 years on a gluten-free diet. Interestingly, for the most part differences with controls were less pronounced than in active disease, suggesting that celiac disease-related methylation changes may be partially reversible or indeed it may be that more time on a gluten-free diet may be necessary to reverse or normalize methylation levels [25]. Additionally histone modifications may play a role in IBD conditions. Treatment with HDACi reduced inflammation in experimental models of colitis via the downregulation of inflammatory mediators and the parallel upregulation of apoptosis markers [26, 27]. HDACi may therefore provide a therapeutic avenue for the spectrum of IBD conditions [28].

As is the case with a number of inflammatory conditions, celiac patients may have an increased risk of cancer development [29, 30]. Two recent studies have provided some evidence to suggest that epigenetic factors may be involved in the development of small bowel adenocarcinoma in celiac patients. Bergmann et al. [31] determined that aberrant CpG island methylation and microsatellite instability link celiac disease and carcinogenesis. Meijer and colleagues found that higher rates of promoter hypermethylation of the *APC* gene were observed in a cohort of small bowel adenocarcinoma patients: 48 % in non-celiac disease-related versus 73 % in celiac disease-related cases [32]. Elevated rates of microsatellite instability were also detected: 33 % in non-celiac versus 67 % in celiac disease-related small bowel adenocarcinomas [32].

Finally a recent study has shown that the miRNA-449a is overexpressed in the small intestine of pediatric celiac patients. In this study this miRNA was found to target both *NOTCH1* and *KLF4* resulting in their downregulation [33]. However, in other settings another known target of this miRNA is in fact *HDAC1* [34], and this suggests that aberrant miRNAs and epi-miRNAs may play critical roles in celiac disease.

Adverse epigenetic events occur frequently in a number of autoimmune-related conditions such as rheumatoid arthritis, multiple sclerosis, and diabetes to name but a few [35–37]. Although few studies have examined the role of epigenetics in the pathogenesis of celiac disease, a number of studies have demonstrated aberrant epigenetic involvement in other diseases affecting the bowel such as Inflammatory Bowel Disease (IBD) [38]. These studies in IBD indirectly support the notion that aberrant epigenetic regulation of gene expression may be an important element in gut pathologies such as celiac disease, thus warranting further investigation.

Here we describe the initial identification of genes, which respond to epigenetic treatment, through RNA isolation and RT-PCR. We also illustrate an X-ChIP assay to determine the effect of an HDi on the promoter region of genes of interest using PCR with primers designed to the 5′ UTRs (untranslated region) contained within their nucleotide sequences. A flow chart of the ChIP method described is outlined in Fig. 2.

Finally we detail two assays for DNA methylation analysis of specific genes of interest. Both of these techniques are dependent on bisulfite modification of genomic DNA and subsequent PCR amplification. In this reaction, treatment of genomic DNA with sodium bisulfite followed by an alkali deaminates cytosine residues thus converting them to uracil, while 5-mC is protected from this modification [39, 40]. The DNA sequence under investigation is then PCR amplified with primers designed to anneal specifically with bisulfite-converted DNA. This combination of bisulfite treatment and PCR means that all uracil and thymine nucleotides are amplified as thymine, whereas only 5-mC is amplified as cytosine. The first of the two methods presented here, MS-HRM, is a fast, reliable method for methylation screening or discovery. It is based on the principal that following bisulfite conversion and PCR amplification, a methylated amplicon will have a higher GC content and thus higher melting temperature than an unmethylated amplicon of the same sequence [41–43]. By incorporating a fluorescent intercalating dye into the PCR reaction and subjecting the PCR product to a denaturing temperature

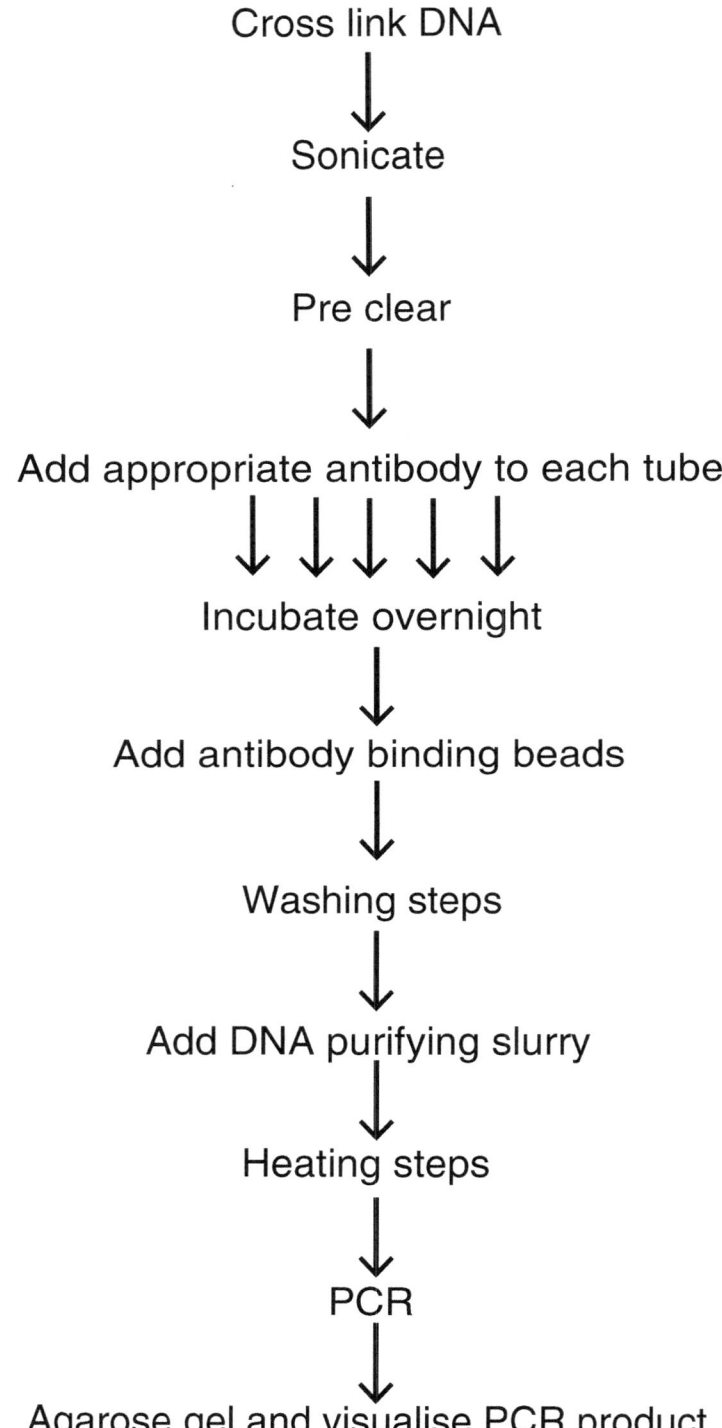

Fig. 2 Overview of the ChIP protocol

gradient, the degree of methylation of a given sequence is calculated by comparing its melting profile with that of known methylation standards (Fig. 3) [44].

In this way, the entire DNA sequence between a primer pair is screened, thus providing a snapshot of the methylation density of multiple CpG sites (as many as are located in a given amplicon). MS-HRM is performed directly after PCR, with no need for cleanup, which makes it a relatively fast technique. The advantage of this inexpensive method is that it can be applied to any genomic region to detect varying degrees of methylation (e.g., partial

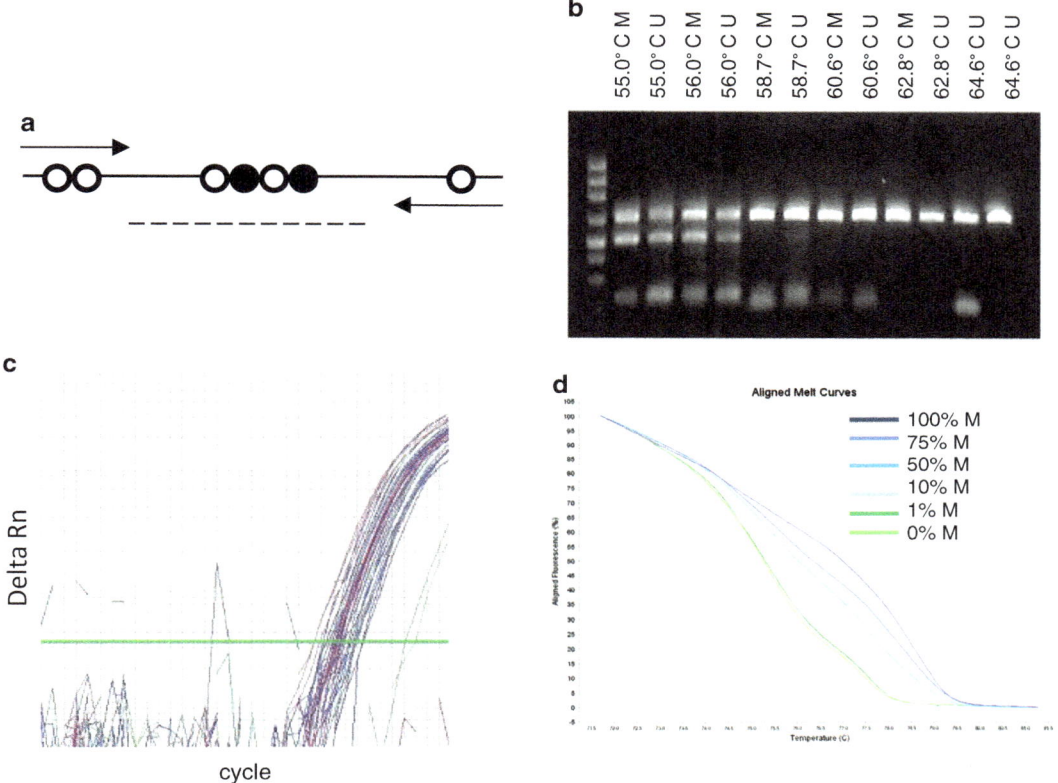

Fig. 3 The principles of HS-HRM for DNA methylation analysis. (**a**) Primers amplify bisulfite-converted DNA, regardless of methylation status. After PCR, differential methylation of the intervening sequence is examined by melt analysis. *Filled circles* indicate 5mC; *white circles* represent C. The *dashed line* indicates the sequence examined by this methodology. (**b**) Annealing temperature gradient reveals the effect of increasing temperature on equal amplification intensities of both methylated (M) and unmethylated (U) DNA. In this case, 60.6 °C is selected as the optimum temperature. (**c**) Analysis of qPCR amplification curves at the optimized annealing temperature should show similar levels of amplification between all samples types. (**d**) HRM melt analysis of methylation standards reveals the sensitivity of detection, in this case 10 %, which is influenced greatly by annealing temperature. (**e**) Overlaying test samples with standards indicates degree of methylation. In this case all test samples are unmethylated

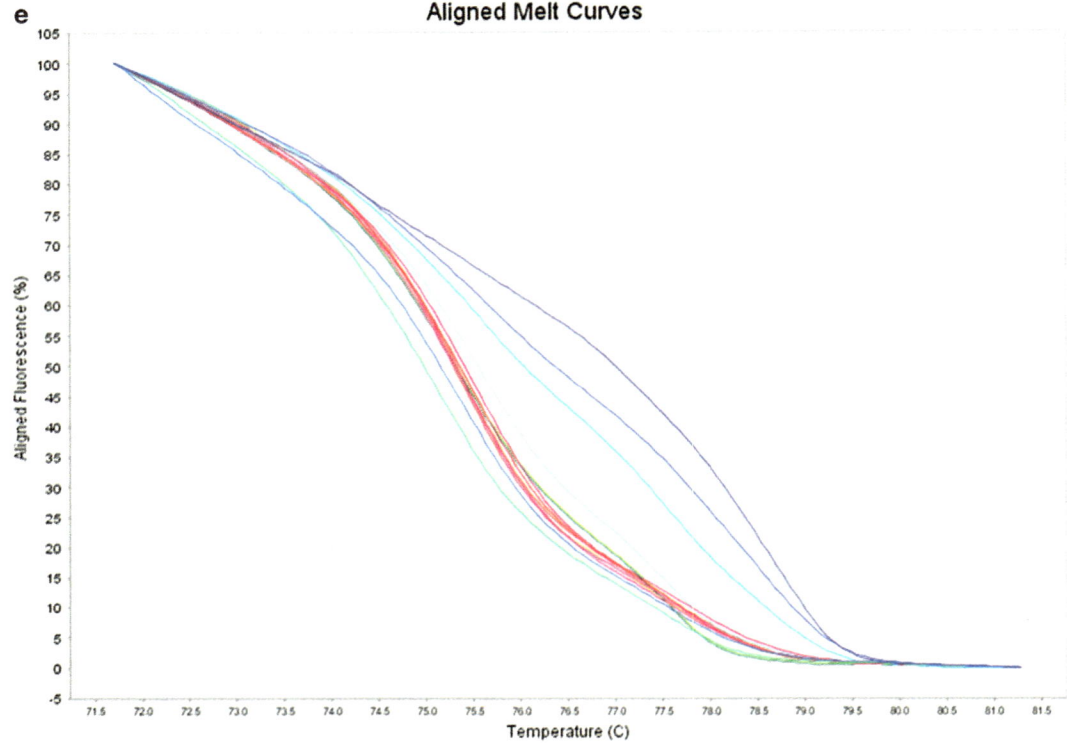

Fig. 3 (continued)

methylation) across a large number of CpG sites. This information can then be used to uncover methylation "hotspots." Sequencing techniques (pyrosequencing, bisulfite sequencing) are needed to reveal the identity of the specific CpGs that are subject to methylation.

The second technique described herein for DNA methylation analysis, qMSP, differs from MS-HRM in that it will only reliably quantify methylation when 100 % of the CpG sites being interrogated by a primer/probe set are methylated. This is achieved by incorporating CpG dinucleotides into the primer and probe binding sites (Fig. 4).

Thus, the PCR amplification will only take place when all of the CpGs in the hybridization sites are represented by 5mC and not uracil. Therefore, if anything, this technique can underestimate the degree of methylation at a given locus. qMSP is especially useful for high-throughput analysis and is extremely sensitive (down to 1/10,000–100,000 methylated alleles in a background of unmethylated alleles) [45].

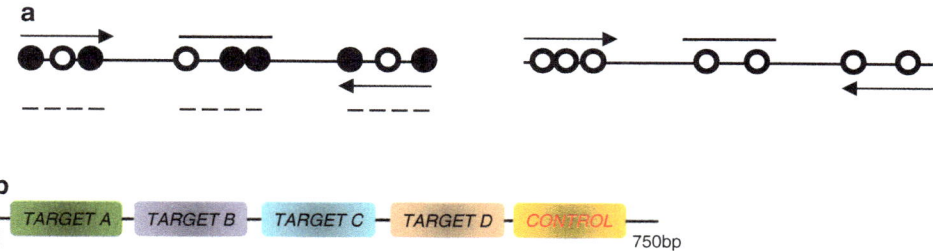

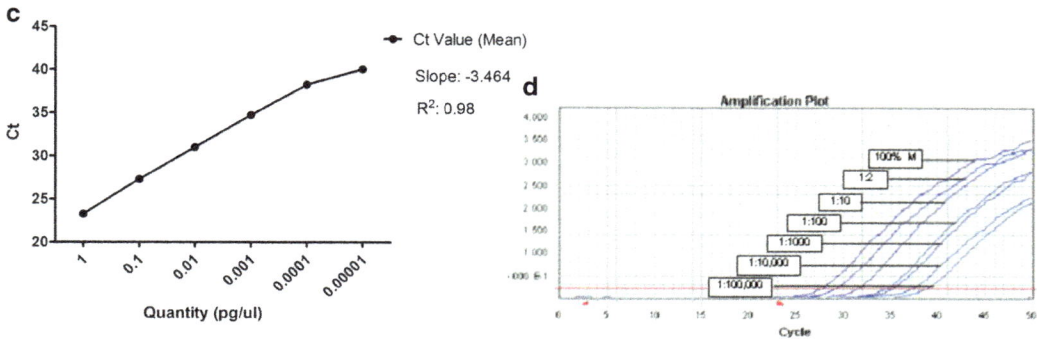

Fig. 4 The principles of qMSP for DNA methylation analysis. (**a**) Primers and fluorescently labelled probe amplify bisulfite-converted fully methylated DNA. A control PCR reaction that does not discriminate between methylated and unmethylated template is used to normalize the amount of input bisulfite modified DNA. *Filled circles* indicate 5mC; *white circles* represent C. The *dashed line* indicates the sequence examined by this methodology. (**b**) Depiction of gBlock design, encompassing multiple target genes and a control for normalization. (**c**) Standard curve results from a qMSP performed on serial dilutions of a gBlock gene fragment. (**d**) Sensitivity of detection of DNA methylation (1/10,000–1/100,000) and quantitative accuracy of qMSP. Amplification plot of fluorescence intensity (*y* axis) against PCR cycle (*x* axis). Each curve represents a different input quantity of in vitro methylated DNA into unmethylated DNA

2   Materials

The reagents and chemicals used should be of analytical grade and stored in accordance with the manufacturers' instructions. Buffers and solutions should be prepared using distilled water and stored at room temperature unless stated otherwise. All local regulations in relation to the handling and disposal of reagents and chemicals must be followed.

2.1   RNA Isolation

1. TRI reagent® (Molecular Research Center, Cincinnati, OH, USA). This reagent allows for the extraction of RNA, DNA, and protein from a single sample, thus saving on time and expense associated with various isolation methods. We consistently obtain a large concentration of pure RNA using this reagent.

2. 1-Bromo-3-chloro-propane (BCP).

3. Isopropanol (2-isopropanol).

4. Diethylpyrocarbonate (DEPC)-treated water.

5. Ethanol (EtOH) wash buffer: 70 % solution in water.

2.2 RQ1 DNAse Treatment

1. RNA qualified (RQ1) RNase-Free DNase.

2. Phenol:Chloroform:Isoamyl Alcohol 25:24:1 Saturated with 10 mM Tris–HCl, pH 8.0, 1 mM EDTA (*see* **Note 1**).

3. Ammonium acetate: 5 M solution in water. Weigh 19.27 g ammonium acetate and dissolve in a small amount of water (*see* **Note 2**). Add water to make up to 50 mL.

4. Polyacryl carrier (Molecular Research Center, Cincinnati, OH, USA) (*see* **Note 3**).

5. 96 % ethanol.

2.3 Complementary DNA Synthesis

1. Oligo(12)$_{20}$ (*see* **Note 4**).

2. 10 mM dNTP stock (*see* **Note 5**).

3. Ribonuclease inhibitor.

4. First strand (FS) buffer.

5. Reverse transcriptase enzyme.

2.4 ChIP Assay

1. High quality 37 % formaldehyde solution.

2. SDS buffer: 10 % solution in water (*see* **Note 6**).

3. Tris–HCl buffer: 100 mM solution in water. Weigh 6.056 g Tris–HCl base and transfer to a graduated cylinder containing approximately 250 mL of water. Make up to 500 mL with water.

4. EDTA buffer: 0.5 M EDTA, pH 8.1. Weigh 146.12 g EDTA and transfer to a 1 L graduated cylinder containing approximately 500 mL of water. Adjust pH to 8.1 with NaOH (*see* **Note 7**). Make up to 1 L with water.

5. Protease inhibitor cocktail (PIC). Store at –20 °C.

6. Phenylmethylsulfonyl fluoride (PMSF). 87 mg/mL solution in 96 % EtOH (*see* **Note 8**). Store at –20 °C.

7. ChIP SDS lysis buffer: 1 % SDS, 10 mM EDTA, and 50 mM Tris–HCl, pH 8.1 (*see* **Note 9**).

8. Appropriate antibodies and controls (*see* **Note 10**).

9. ChIP buffer (5×) (Based on our experience with this assay, we recommend the One-Day ChIP kit from Diagenode, Liege, Belgium).

10. Deionized water.

11. Protease Inhibitor mix (200×) (Diagenode, Liege, Belgium).

12. Antibody-binding beads (Diagenode, Liege, Belgium).

13. DNA-purifying slurry (Diagenode, Liege, Belgium).

14. Proteinase K (Diagenode, Liege, Belgium).

15. Microcentrifuge tube mini floating rack.

2.5 RT-PCR Setup and Agarose Gel Electrophoresis

1. GoTaq® Green Master Mix (400 μM dATP, dGTP, dCTP, dTTP, 3 mM $MgCl_2$, GoTAQ, pH 8.5) (Promega, Madison, WI, USA): In our hands we obtain excellent results with this master mix, but other master mixes from different vendors can also be used.

2. 50× TAE buffer: 2 M Tris–HCl, 50 mM EDTA, pH 8.0. Weigh 242 g Tris–HCl base and transfer to a 1 L glass bottle containing approximately 500 mL of water. Add 57.1 mL glacial acetic acid and 100 mL 0.5 M EDTA (pH 8.0) buffer. Place a magnetic stir bar (flea) into the bottle and place atop of a magnetic stirrer on a medium speed for approximately 30 min until all of the Tris–HCl base has dissolved. Transfer into a graduated cylinder and make up to 1 L with water (*see* **Note 11**).

3. DNA ladder.

4. Electrophoresis set.

2.6 Methylation-Sensitive High Resolution Melt Analysis (MS-HRM)

1. Bisulfite modification kit (e.g., EpiTect Fast Bisulfite Conversion kit, Qiagen EZ DNA methylation-lightening kit, Zymo Research).

2. HRM master mix containing Taq polymerase and fully saturating dye (e.g., MeltDoctor HRM master mix, Life Technologies, EpiTect HRM PCR kit, Qiagen).

3. $MgCl_2$.

4. Control human DNA for constructing standard curves: 100 % methylated, bisulfite modified, 100 % unmethylated, bisulfite modified and unmodified genomic DNA (e.g., EpiTect PCR control DNA set, Qiagen).

5. DNA primers (standard desalting purified), e.g., Integrated DNA Technologies.

6. 0.2 mL PCR strip tubes.

7. 96-well PCR plates (e.g., ABgene FAST 96-well PCR plate, Thermo Scientific, MicroAmp[R] Fast optical 96-well reaction plates, Life Technologies).

8. Optical adhesive covers (e.g., Thermo Scientific, MicroAmp[R] Optical adhesive covers, Life Technologies).

9. Real-Time PCR instrument with HRM capabilities (e.g., Applied Biosystems 7900HT Fast Real-Time PCR system with SDS software v2.3 or later, Qiagen Rotor-Gene, and Roche LightCycler 480 Real-Time PCR system).

2.7 Quantitative Methylation Specific PCR (qMSP)

1. Bisulfite modification kit (as in Subheading 2.6).

2. qPCR master mix without AmpErase® Uracil N-Glycosylase (e.g., TaqMan® Universal PCR Master Mix, No AmpErase® UNG, Life Technologies).

3. 96-well PCR plates (e.g., ABgene FAST 96-well PCR plate, Thermo Scientific, MicroAmp^R Fast optical 96-well reaction plates, Life Technologies).

4. Optical adhesive covers (e.g., Thermo Scientific, MicroAmp^R Optical adhesive covers, Life Technologies).

5. DNA primers (standard desalting purified) and fluorescently labelled probe (e.g., Zen double-quenched probes, Integrated DNA Technologies, MGB-quenched probes, Life Technologies).

6. gBlocks Gene Fragments for constructing standard curves and interplate calibration (e.g., Integrated DNA Technologies).

7. Control Human methylated DNA for relative quantification of methylation (e.g., EpiTect Control methylated DNA, Qiagen).

3 Methods

Care must be taken when working with RNA. Designate a special area for RNA work only. Treat all surfaces with an RNase inactivating agent, use RNase-free plastic and DEPC-treated water.

Expose your cell lines of interest to an epigenetic targeted agent such as an HDACi for an appropriate period of time. It is common to treat with an HDACi for a period of 24 h. Proceed to Subheading 3.1. Once you have determined that your gene of interest is responsive to epigenetic agent(s), as judged by increased levels of PCR product in your treated versus untreated sample, continue to Subheading 3.4.

3.1 RNA Isolation

1. When cell line(s) have been treated with the appropriate drug for the recommended time frame, decant medium from tissue culture flask and add 1 mL TRI reagent®. Incubate with agitation on a shaker for 5 min.

2. Scrape cells from the bottom of the flask using a cell scraper. Transfer TRI reagent® to a 1.5 mL microfuge tube.

3. Add 100 μL BCP to each sample, invert for 15 s, and incubate for 10 min. Centrifuge samples at $13,500 \times g$ for 15 min.

4. Transfer colorless upper aqueous phase (containing RNA) to a clean 1.5 mL microfuge tube and discard the remaining phases (*see* **Note 12**).

5. Add 500 μL isopropanol (2-propanol) to each sample. Mix by inversion and incubate for 10 min. Centrifuge at $13,500 \times g$ for 8 min.

6. Decant supernatant and add 1 mL EtOH wash buffer to RNA pellet. Incubate for 5 min.

7. Centrifuge at $13,500 \times g$ for 5 min and decant EtOH wash. Air-dry pellet for 5 min and resuspend in 50 µL water (*see* **Note 13**).

8. Quantify the RNA (*see* **Note 14**). Proceed to Subheading 3.2 or store at −80 °C until required.

3.2 RQ1 DNase Treatment

1. Transfer 1 µg RNA to a microfuge tube. Add 5 µL 10× RQ1 buffer, 1 µL RQ1 DNase to the RNA sample and adjust volume to 50 µL with water (*see* **Note 15**). Mix samples with a pipette tip and incubate at 37 °C for 1.5 h.

2. Add 50 µL phenol-chloroform to each sample. Vortex briefly and centrifuge at $13,500 \times g$ for 5 min.

3. Transfer the upper layer containing the RNA to a clean microfuge tube. Add 1 µL polyacryl carrier, 50 µL 5 M ammonium acetate, and 300 µL 96 % EtOH. Vortex and incubate for 15 min.

4. Centrifuge at $13,500 \times g$ for 20 min. Discard supernatant and wash pellet with 1 mL ethanol wash buffer (70 %). Centrifuge at $13,500 \times g$ for 5 min.

5. Discard the EtOH wash, resuspend pellet in 10 µL water. Proceed to Subheading 3.3 or store at −80 °C until required.

3.3 cDNA Preparation

1. Add 1 µL 50 µM Oligo(dT)$_{20}$, 1 µL 10 mM dNTP stock, and 1 µL water to the 10 µL RNA sample from Subheading 3.2 (final reaction volume 13 µL). Mix sample with a pipette.

2. Incubate at 65 °C for 5 min and transfer to ice for 1 min.

3. Add 4 µL 5× FS Buffer, 1 µL 0.1 M DTT, 1 µL ribonuclease inhibitor (40 U/µL), and 1 µL reverse transcriptase enzyme (200 U/µL) to each sample. Mix gently with a pipette and incubate at the vendors recommended temperature for 1 h. Stop the reaction by incubating the samples at 70 °C for 15 min (*see* **Note 16**). Proceed to Subheading 3.6 or store samples at −20 °C until required.

3.4 Chromatin Immunoprecipitation Assay (X-ChIP)

When cells have been treated with a suitable histone deacetylase inhibitor (HDACi) or DNA methyltransferase inhibitor (DNMTi) with corresponding vehicle control (untreated) for the appropriate period of time, the X-ChIP assay is carried out over 2 days. Volumes indicated here are for 20 immunoprecipitates (*see* **Note 17**), and are based on the One-Day ChIP kit from Diagenode, which is our preferred X-ChIP kit. Scale volumes accordingly for different sample numbers. On day 1 the DNA is cross-linked, sheared, and incubated with appropriate antibody overnight while on day 2, the immunoprecipitation is completed and DNA purification steps are performed. The result is a sample of DNA ready for PCR. All centrifugation steps should be performed at 4 °C unless stated otherwise.

Day 1

1. Add formaldehyde directly to the cell culture medium within the tissue culture flask to a final concentration of 1 % (270 μL in 10 mL of culture medium). Mix thoroughly and incubate for 5 min at 37 °C. Decant medium and wash cells twice with ice-cold PBS.

2. Add 1 mL ice-cold PBS into the flask and scrape cells using a cell scraper and transfer into a clean 1.5 mL microfuge tube. Centrifuge samples at 14,000×*g* for 4 min.

3. Remove supernatant and resuspend pellet in 200 μL ChIP SDS lysis buffer (*see* **Note 18**). Incubate on ice for 30 min.

4. Sonicate samples, while keeping on ice. Sonicate for a period of 10 s, followed by a rest of 30 s (*see* **Note 19**). Repeat three times.

5. Centrifuge for 10 min at 13,000×*g*. Transfer 200 μL supernatant (sheared DNA) into a clean microfuge tube. Remove 10 μL supernatant for analysis (*see* **Note 20**).

6. Thaw appropriate antibodies and Protease Inhibitor Cocktail (PIC) on ice.

7. Prepare 1× ChIP buffer (*see* **Note 21**). Dilute the remaining sheared DNA tenfold (190 μL supernatant and 1.9 mL of buffer).

8. Dilute sheared DNA to a maximum of 28 % sheared DNA relative to 1× ChIP/protease inhibitor buffer volume (e.g., 1.32 mL sheared DNA: 4.84 mL buffer) (*see* **Note 22**).

9. Transfer a 280 μL aliquot of the diluted sheared DNA into appropriately labelled 1.5 mL microfuge tubes (on ice) (*see* **Note 23**). Transfer 6 μL into a separate microfuge tube; this will serve as the Input DNA (*see* **Note 24**) in later steps. Freeze excess at −80 °C.

10. Add antibody to appropriately labelled tubes (*see* **Note 25**) leaving one set of tubes free of antibody to serve as the no antibody control (*see* **Note 26**) and vortex briefly. Incubate overnight at 4 °C on a rotary shaker.

Day 2

1. Resuspend antibody-binding beads and transfer 840 μL into 10.5 mL 1× ChIP buffer. Mix by inversion and centrifuge for 3 min at 500×*g*. Remove supernatant and resuspend beads in 10.5 mL 1× ChIP buffer. This is enough for 20 individual IP reactions.

2. Transfer 500 μL of the washed beads into 1.5 mL microfuge tubes (*see* **Note 27**). Centrifuge for 2 min at 500×*g*. Check by eye that equal pellets of beads are present in each tube (*see* **Note 28**). Remove supernatant. Store tubes on ice until required.

3. Centrifuge microfuge tubes containing antibody (from overnight incubation, Day 1, **step 10**) for 10 min at 12,000×*g*.

4. Transfer 250 μL each supernatant (*see* **Note 29**) to the pelleted beads (derived from **step 2** above). Incubate for 30 min at 4 °C on a rotary shaker.

5. Thaw Input DNA aliquot (from Day 1, **step 9**) and add 30 μL 96 % EtOH. Incubate on ice for 10 min and centrifuge at 10,000×*g* for 10 min. Discard supernatant and briefly allow pellet to air-dry. Resuspend pellet in 100 μL water and incubate at room temperature until **step 10** below.

6. Add 1 mL ice-cold 1× ChIP buffer to all tubes from **step 4**. Invert twice. Centrifuge at 500×*g* for 2 min.

7. Remove the supernatant and resuspend beads in 1 mL ice-cold 1× ChIP buffer. Transfer to a fresh 15 mL tube containing 12 mL cold ChIP buffer (*see* **Note 30**). Incubate at 4 °C for 5 min and centrifuge for 3 min at 500×*g* at 4 °C.

8. Remove 12 mL buffer from each tube and resuspend beads in the 1 mL remaining buffer. Transfer to fresh 1.5 mL microtube (on ice). Centrifuge for 2 min at 500×*g*.

9. Remove supernatant. Allow pellets to air-dry.

10. Resuspend the DNA-purifying slurry and add 100 μL to each bead pellet and the Input DNA sample from **step 5**. Incubate for 1 min. Invert tubes and incubate for 10 min in 100 °C water (*see* **Note 31**).

11. Cool samples and add 1 μL Proteinase K to each and vortex. Incubate for 30 min at 1000 revolutions per minute (rpm) in a thermomixer at 55 °C.

12. Incubate in 100 °C water for 10 min (*see* **Note 31**). Centrifuge at 14,000×*g*. The resulting supernatant is ready for subsequent PCR reactions (*see* **Note 32**).

3.5 *Primer Design*   ChIP primers should be designed towards the promoter region of your gene of interest from the known promoters and/or 5′ UTRs contained within their nucleotide sequences (*see* **Note 33**). Standard cycling conditions: Template DNA is initially denatured at 94 °C for 5 min, followed by 35 amplification cycles in a thermal cycler. Each cycle consists of template denaturation (1 min at 94 °C), primer annealing (1 min at target annealing temperature), and extension (1 min at 72 °C). This is followed by an elongation step (72 °C for 10 min) to complete the amplification cycle (*see* **Note 34**).

3.6 Reverse Transcriptase Polymerase Chain Reaction (RT-PCR)

cDNA (Subheading 3.3) or DNA from the ChIP assay (Subheading 3.4) is used as a template for the polymerase chain reaction (PCR).

1. Prepare the following PCR master mix for each sample to be amplified: 10 μL 2× GoTaq® Green Master Mix, 2 μL 5 μM forward primer, 2 μL 5 μM reverse primer, and 6 μL water (final reaction volume 20 μL).

2. Use 1 μL cDNA template per reaction (2.5 μL for ChIP and adjust water accordingly). Include a negative control for each RT-PCR reaction (*see* **Note 35**).

3.7 Agarose Gel Electrophoresis

All PCR products are resolved by electrophoresis on an agarose gel (*see* **Note 36**) containing ethidium bromide (*see* **Note 37**) and visualized under a UV gel system (*see* **Note 38**).

1. Weigh 2 g agarose and dissolve in 100 mL 1× TAE by boiling in a microwave for 2–3 min on a medium heat setting. Check periodically and swirl repeatedly.

2. Allow solution to cool to 55–60 °C (hand hot), before the addition of ethidium bromide, to achieve a final concentration of 1 μg/mL.

3. Pour the 2 % agarose solution into a gel tray with well-forming combs to a depth of 3–5 mm and allow to solidify (approximately 20–30 min). Once solidified pour sufficient 1× TAE buffer to the maximum level as indicated on gel rig. Remove combs.

4. Load a 10 μL aliquot of each PCR product (including negative) to individual wells. Load an appropriate DNA ladder. Keep voltage constant and allow electrophoresis to run for approximately 30–40 min (depending on voltage) until dye front is approximately ¾ of the way down the gel (*see* **Note 39**).

5. Figure 5a, b is provided for illustration purposes.

Interestingly, the IL-23 axis has previously been shown to play important roles in celiac disease [46–48].

3.8 Primer Design for DNA Methylation Analysis

1. CpG islands and corresponding genomic DNA sequence can be freely viewed and downloaded at the UCSC Human Genome Browser: http://genome.ucsc.edu/

2. Copy genomic region of interest into a word document and transform it into a bisulfite modified fully methylated sequence. Do this by first making sure all of the sequence is in lower case. Use the Find and Replace function to replace all "cg" with "CG". Next, replace all "c" with "t", taking care to ensure that "match case" option is selected. Use this sequence for primer design.

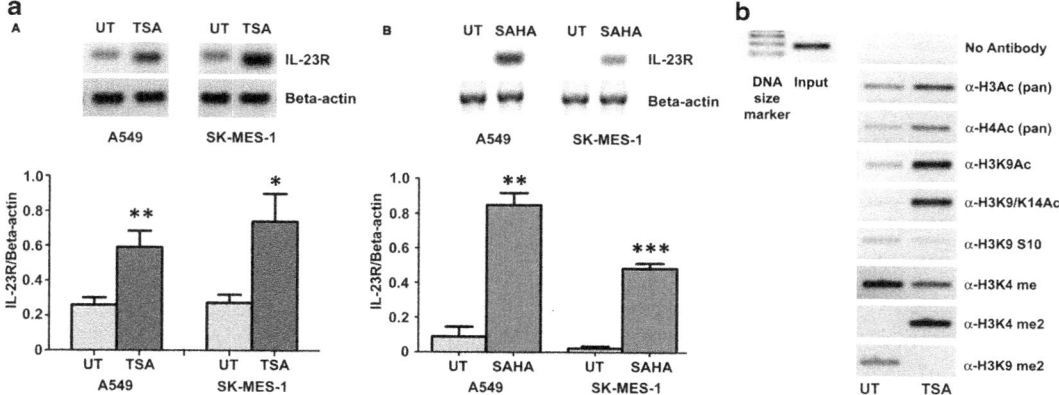

Fig. 5 An example of X-Chip. (**a**) The effect of (*a*) TSA and (*b*) SAHA treatment on the expression of *IL-23R* on A549 and SK-MES-1 cells. Cells were treated for 24 h with TSA at a final concentration of 250 ng/mL and with SAHA at a final concentration of 5 μM. Densitometry analysis of expression in treated versus untreated samples when normalized to beta actin. Data is graphed as mean ± standard error of mean (*n* = 3). Statistical analysis performed using a one tailed student's *t* test (*$p < 0.05$; **$p < 0.01$; ***$p < 0.001$). (*UT* untreated, *TSA* Trichostatin A, *SAHA* suberoylanilidehydroxamic acid). (**b**) Histone acetylation occurs directly at the promoter region of *IL-23R* in the A549 cells after treatment with TSA (250 ng/mL for 24 h). As evidenced by increased PCR product for the various markers within the TSA-treated samples. The ChIP assay was performed using the following antibodies: pan acetyl-histone H3 (H3Ac), pan acetyl-histone H4 (H4Ac), acetyl-histone H3 Lys 9 (H3K9Ac), acetyl-histone H3 Lys 9/14 (H3K9/14ac), acetyl-histone H3 Lys 9 phosphoSer10 (H3K9S10), methyl-histone H3 Lys 4 (H3K4Me), di methyl-histone H3 Lys 4 (H3K4Me2), and di methyl-histone H3 Lys 9 (H3K9Me2). Input DNA serves as a positive control. A no antibody control was included to test for nonspecific binding. (*UT* untreated, *TSA* Trichostatin A). Figures as originally published in Baird et al. [51]

3. Oligonucleotides (primers and probes) should contain several non-CpG cytosines, which appear as thymines in the in silico modified sequence, to ensure amplification only of bisulfite modified DNA.

4. Primers should meet standard parameters for primer design, e.g., avoid secondary structures, self-dimers and hetero-dimers. Ensure that the melting temperature of the primers is matched, preferably within 1 °C and typically between 58 and 60 °C. We recommend using the freely available Oligo Analyzer from Integrated DNA technologies (http://eu.idtdna.com/ analyzer/Applications/OligoAnalyzer/), or the UCSC in silico PCR platform (http://www.genome.ucsc.edu/).

5. To design primers for MS-HRM, the primers should amplify bisulfite modified DNA regardless of methylation status (Fig. 3a). Therefore, in addition to **steps 3** and **4**, CpG sites should be avoided in the primer sequences. However, a single CpG site can be included at the 5′ end to avoid PCR bias of unmethylated templates. The amplicon length should be <300 bp to reduce complexity of the melting profile.

The number of CpG sites within the amplicon may also be an important factor to consider.

6. To design primers for qMSP, the primers and probe should only amplify bisulfite modified methylated DNA (Fig. 4a). Therefore, in addition to **steps 3** and **4**, at least two CpG sites should be incorporated into each oligonucleotide, preferably towards the 3′ end. In addition, a control reaction is performed with primers that will only amplify bisulfite modified DNA, regardless of DNA methylation, in order to normalize for varying amounts of bisulfite modified DNA between samples. Control oligonucleotides should avoid CpG sites but must contain several non-CpG cytosines. The amplicon length should be <150 bp and the melting temperature of the probe should be approximately 10 °C greater than the primers to comply with standard real-time PCR parameters.

3.9   MS-HRM

1. Isolate DNA from starting source (e.g., cell line, tissue sample) using a method of choice.

2. Perform bisulfite conversion of DNA. Many commercially available kits are optimized to modify as little as 100 pg up to 2 μg DNA. In our hands, this technique performs best using 100–500 ng of input genomic DNA. Take care at the final elution step to avoid over-concentrating the bisulfite-converted sample. We recommend eluting into a final volume that yields a concentration in the region of 10 ng/μL, e.g., 500 ng of input genomic DNA eluted into a final volume of 50 μL, thus providing sufficient volume of converted DNA for multiple PCR reactions, as required. There are no methods to specifically quantify bisulfite modified DNA. Therefore, calculations of concentration are based on the assumption of 100 % conversion rate of the reaction.

3. Prepare a set of seven methylation controls by serially diluting 100 % methylated DNA into 100 % unmethylated DNA as follows: *100 %, 75 %, 50 %, 10 %, 1 %, 0.1 %, 0 %* (Table 1).

   Keep the total DNA quantity constant between the dilutions. Therefore, each of the seven standards should amplify with comparable CT values.

4. Prepare MS-HRM PCR reactions as follows: HRM PCR buffer containing Taq polymerase and saturating dye, 300 nM final concentration of both forward and reverse primers, 10 ng of bisulfite modified DNA and H_2O to bring total reaction volume to 20 μL, *see* **Note 40**. Centrifuge PCR plates briefly at $1000 \times g$ before placing into the instrument. For instructions on MS-HRM PCR optimization, *see* **Note 41**.

5. Perform the PCR reaction and melting analysis in a Real-Time instrument (*see* Subheading 2.6) under the following thermal

Table 1
Preparation of methylation standards for MS-HRM

Standard	Methylation %	Methylated DNA (10 ng/µL) (µL)	Unmethylated DNA (10 ng/µL) (µL)	Total DNA (ng)
1	100	20	0	200
2	75	15	5	200
3	50	10	10	200
4	10	2.2	19.8	220
5	1	2.2 (standard 4)	19.8	220
6	0.1	2 (standard 5)	18	200
7	0	0	20	200

200 ng of DNA should provide enough material for 20 individual PCR reactions

cycling conditions: an initial hot-start at 95 °C for 10 min followed by 40 cycles of 95 °C for 15 s, assay-specific annealing temperature (typically 58–68 °C) for 30 s and 72 °C for 30 s, concluding with a final extension at 72 °C for 5 min.

6. Perform the melt curve/dissociation immediately after amplification by denaturing PCR products at 95 °C and allowing them to re-anneal at 60 °C, according to the manufacturer's guidelines. Ramp rate may need to be manually adjusted from the standard 100 to 1 % during the denaturing/melting step.

7. Analyze HRM data with appropriate instrument software (e.g., HRM Software v3.0.1 software, Life Technologies) (Fig. 3d, e).

3.10 qMSP

1. Isolate DNA and perform bisulfite modification as described in Subheading 3.9.

2. Prepare tenfold serial dilutions of a bisulfite modified methylated DNA to construct a standard curve for Absolute Quantification (AQ), *see* **Note 42** and Table 2.

It is essential that the concentrations of the standards are such that their amplification range spans that of the unknowns or samples to be measured.

3. For each DNA sample and standard, a parallel real-time PCR reaction must be performed with oligonucleotides targeted to a control gene (e.g., *ACTB*) to normalize for the amount of input bisulfite modified DNA between samples (*see* Subheading 3.8 and Fig. 4a).

4. Prepare qMSP reactions as follows: qPCR master mix without AmpErase® Uracil *N*-Glycosylase, 300–900 nM final concentration of both forward and reverse primers, 100–300 nM final

Table 2
Preparation of methylation standards for qMSP

Standard	Copy number	Volume of gBlock™ (µL)	Volume of molecular grade H$_2$O (µL)
1	1,000,000	10 (WS)	173.8
2	100,000	10 (1)	90
3	10,000	10 (2)	90
4	1000	10 (3)	90
5	100	10 (4)	90
6	10	10 (5)	90

We suggest preparing a working solution (WS) of the gBlock™ at a concentration of 10 pg/µL

concentration of fluorescently labelled probe, 10 ng of bisulfite modified DNA and H$_2$O to bring total reaction volume to 20 µL. For details on primer/probe optimization, *see* **Note 43**.

5. Perform all reactions in triplicate; this should be factored in when calculating volumes for a PCR master mix.

6. Each target gene being quantified (for methylation levels) must be amplified in the study samples, the methylation standards (minimum 4), a positive control (fully methylated human DNA), and a negative template control.

7. Perform qMSP under standard Absolute Quantification (AQ) real-time settings, adjusting to 50 cycles of amplification.

8. Examine the amplification of the controls and standards first. Adjust the threshold and baseline if necessary so that the slope of the standards is −3.3 (±0.2) and the R^2 falls within the range of 0.997–0.999.

9. The software will automatically extrapolate from the standard curve to give quantities (ng) of methylation for each unknown/sample.

10. Analyze qMSP data by calculating a normalized index of methylation (NIM) for each sample, as previously described [49, 50]. This will determine the ratio of the normalized amount of methylated target gene to the normalized amount of control gene, by applying the formula:

$$\left[\left(\text{TARGET}_{\text{sample}} / \text{TARGET}_{\text{MC}}\right) / \left(\text{CONTROL}_{\text{sample}} / \text{CONTROL}_{\text{MC}}\right)\right] \times 1000$$

where TARGET$_{sample}$ is the quantity of fully methylated copies of a gene of interest in any individual sample, TARGET$_{MC}$ is the quantity of fully methylated copies of a gene of interest in the methylated control DNA, CONTROL$_{sample}$ is the quantity of bisulfite modified templates in any individual sample, and CONTROL$_{MC}$ is the quantity of bisulfite modified templates in the universally methylated control DNA.

4 Notes

1. When taking an aliquot, make sure to go below the upper layer into the organic phase.

2. Having a small amount of the appropriate solvent at the bottom of a bottle or graduated cylinder helps to dissolve most chemicals quicker. Make sure the entire chemical has dissolved before making the solution up to its final volume.

3. PolyAcryl carrier intercalates with RNA, allowing for easier identification of the RNA pellet. The addition of this carrier can affect the optical density of RNA. To normalize for this effect, process a blank sample containing only the reagent (in this case water) and PolyAcryl Carrier, when quantifying RNA samples.

4. We use Oligo(dT)$_{20}$ (homogenous mixture of 20-mer thymidines) for cDNA synthesis as it has a great specificity for mRNA and allows many different targets to be studied from the same cDNA pool.

5. dTTP, dATP, dGTP, and dCTP are usually supplied as individual 100 mM stocks. To make a 10 mM dNTP stock, add 10 μL of each to 60 μL of water (final volume 100 μL).

6. SDS can precipitate out of solution at colder temperatures. Therefore the buffer must be pre-warmed before use.

7. NaOH is essential for the solubility of EDTA. We generally make a dilute (fine-tuning the pH) and concentrated (for large changes) NaOH solution to adjust the pH. We also keep the EDTA solution stirring on a magnetic stirrer to ensure uniform distribution of the NaOH when pH'ing.

8. PMSF is a supersaturated solution that requires vigorous vortexing.

9. A simple way to prepare lysis buffer: Measure 5 mL of 10 % SDS buffer, 5 mL of 100 mM Tris–HCl buffer, and 5 mL of 0.5 M EDTA buffer into a graduated cylinder. Make up to 50 mL with water.

10. Decide which targets are going to be investigated such as acetylated Histone H3 and source ChIP appropriate antibodies.

The negative controls can consist of an IgG antibody raised in the same species as the ChIP antibodies or a no antibody control to detect for nonspecific binding.

11. It is more convenient to make a 50× stock of Tris-Acetate-EDTA (TAE) buffer and dilute to 1× as required (20 mL of 50× stock in 1 L of water). The 0.5 M EDTA buffer is made as mentioned previously with the pH adjusted to pH 8.0 with NaOH.

12. TRIreagent® can be used for the isolation of RNA, DNA, and protein from the one sample. After the initial centrifugation step (subsequent to the addition of BCP), the sample separates into three phases. The upper aqueous phase contains the RNA, while the interphase (middle) contains the DNA and the organic lower phase contains the protein. It is better to leave some of the colorless phase behind to avoid contaminating the RNA.

13. The volume of molecular grade water will vary depending on the size of the RNA pellet. For smaller pellets use about 20 μL water.

14. The 260:280 purity ratio should be approximately 2. If this ratio is lower, it may be due to protein and/or phenol contamination.

15. A master mix can be prepared (buffer, RQ1) for all of the reactions. Always make enough master mix for one or two additional samples to control for pipette error. Master mixes can also be prepared for the cDNA and RT-PCR protocols.

16. The temperature and incubation times will vary depending on the type of reverse transcriptase enzyme that is used. The temperatures and times given here refer to Superscript™ III RT (Invitrogen Corp., Carlsbad, CA, USA).

17. Twenty immunoprecipitates include samples from two treatments (e.g., untreated, HDACi), incubated with eight antibodies, and a no antibody control, then a positive control and an extra sample to control for pipetting error.

18. Warm ChIP SDS lysis buffer (e.g., Millipore) to room temperature before use and add the following (volumes given per 1 mL of buffer): 10 μL Protease Inhibitor Cocktail, 2 μL PMSF.

19. Sonicate samples (to shear DNA) on a medium setting (refer to manufacturers' instructions) by placing needle into the tube just above the bottom. Be careful to avoid contact of the needle with tube surface. Allow sample adequate time to cool between sonications.

20. Run sample on an agarose gel to determine shearing efficiency. If the DNA is inadequately sheared, repeat the sonication

steps. The duration and the number of sonication cycles must be optimized for each cell line before continuing with the rest of the ChIP protocol.

21. Dilute ChIP buffer (5×) to a 1× stock in deionized water (add 100 mL ChIP buffer to 400 mL water). Store at 4 °C.

22. For example for one IP—66 μL sheared DNA: 242 μL buffer.

23. For example if you have an untreated and a treated sample and are testing two antibodies, you will need four tubes, plus tubes for controls.

24. Input DNA serves as a positive control and will be used at a later stage in the protocol.

25. You will need to use approximately 1–2 μg antibody. We have successfully used 2 μg of antibody in these experiments.

26. A no antibody control is used to test for nonspecific binding. These tubes will be treated in the same way as the others, but will not contain any antibody.

27. The beads must be kept in suspension. Mix beads in between each aliquot. Use a tip that has been cut at an angle.

28. Aspirate supernatant slowly. Do not disturb the pellet. If pellet is disturbed, centrifuge the samples again. It is important to check if any buffer remains in the cap and if present remove.

29. Do not disturb the pellet, as this contains the unspecific aggregates. Do not add all of the supernatant antibody-chromatin mix to the pellet of beads. Do not take more than 250 μL from the tubes and discard the excess.

30. Pipette beads against the side of the tube at the top of the liquid as this helps to wash beads further as they sediment through the liquid.

31. Many different cap locks are available and they prevent the cap from popping open due to the heat.

32. Additional information on the Diagenode One-day ChIP Kit protocol can be found at http://www.diagenode.com/media/catalog/file/OneDay_ChIP-Kit_manual.pdf.

33. A number of different software packages are available including free online versions to help with primer design.

34. It is helpful to use a positive control sample to optimize PCR cycling conditions. Ensure that one clean band is present at the expected product size.

35. It is imperative to include a negative sample with each PCR performed. This should consist of all the components of the PCR reaction with water in place of template DNA. No band should appear in the negative control well.

36. The percentage of agarose gel used depends on the size of the PCR product. Use a 2 % gel (2 g agarose in 100 mL 1× TAE) for products below 1 kB.

37. Ethidium bromide is a carcinogenic agent and requires careful handling and proper disposal. Compliance with all local rules regarding its use is mandatory.

38. PCR products were visualized and photographed under UV light using a Biospectrum Imaging System (Ultra Violet Products, Cambridge, UK) in our laboratory.

39. We load sample (e.g., IL-23R) (10 μL) and control (beta actin) (2 μL) side by side on the one gel. Densitometry analysis was performed using the TINA 2.09c densitometry programme (Raytest, Germany). The target gene expression is normalized to the control gene and expressed as ratio of target expression:control gene expression.

40. Life Technologies MeltDoctor™ HRM master mix is in our hands the most robust performing HRM master mix, requiring the least amount of PCR optimization across different temperatures and producing the most consistent amplification curves.

41. It is recommended when performing MS-HRM with a new set of primers that the following steps are taken to ensure optimum PCR amplification of all templates, regardless of degree of methylation. A PCR annealing temperature gradient should be performed from 55 to 65 °C on both 100 % methylated and 100 % unmethylated DNA using a gradient thermal cycler. This can be done in 0.2 mL strip tubes if preferred. Visualization of PCR products by gel electrophoresis will help to select the optimum annealing temperature that generates PCR products of equal intensity on both methylated and unmethylated DNA (Fig. 3b). However, if a CpG site has been incorporated into the 5′ of the primer(s), preferential amplification of the unmethylated template may appear. When the annealing temperature has been selected, PCR and HRM can be performed in a single reaction on a Real-Time Instrument. However, visual inspection of the amplification curves should always be carried out to ensure amplification at similar cycle numbers for all templates.

42. For constructing standard curves for qMSP, we recommend using synthetic ds DNA fragments such as gBlocks™ (Integrated DNA Technologies), which can be in silico engineered as described in Subheading 3.8. Alternatively, commercially available bisulfite modified methylated DNA can be used. gBlocks™ gene fragments have capacity up to 750 bp and can thus be designed to house multiple target genes of interest and a control gene for normalizing input amounts of bisulfite

modified DNA between samples (Fig. 4b). Prepare a working solution of the gBlocks™ at a concentration of 10 pg/µL and use this to prepare methylation standards (Table 2).

43. The primer and probe concentration is assay dependent and needs to be optimized across a range, typically 300, 600, and 900 nM for primers and 100, 200, and 300 nM for the probe. Amplification curves should be visualized to assess the cycle number of amplification and the height of the change in fluorescence emitted. Optimized primer/probe sets must be used on fully methylated bisulfite-converted DNA, fully unmethylated bisulfite modified DNA, and genomic DNA, to ensure specific amplification of bisulfite modified methylated DNA only, before proceeding to sample analysis.

Acknowledgements

The authors gratefully acknowledge the contribution of Alexandra Tuzova in the design of Fig. 1.

References

1. Waddington CH (2012) The epigenotype. 1942. Int J Epidemiol 41(1):10–13. doi:10.1093/ije/dyr184

2. Cyr AR, Domann FE (2011) The redox basis of epigenetic modifications: from mechanisms to functional consequences. Antioxid Redox Signal 15(2):551–589. doi:10.1089/ars.2010.3492

3. Denis H, Ndlovu MN, Fuks F (2011) Regulation of mammalian DNA methyltransferases: a route to new mechanisms. EMBO Rep 12(7):647–656. doi:10.1038/embor.2011.110

4. Portela A, Esteller M (2010) Epigenetic modifications and human disease. Nat Biotechnol 28(10):1057–1068. doi:10.1038/nbt.1685

5. Skene PJ, Henikoff S (2013) Histone variants in pluripotency and disease. Development 140(12):2513–2524. doi:10.1242/dev.091439

6. Dawson MA, Kouzarides T (2012) Cancer epigenetics: from mechanism to therapy. Cell 150(1):12–27. doi:10.1016/j.cell.2012.06.013

7. Bird AP (1986) CpG-rich islands and the function of DNA methylation. Nature 321(6067):209–213. doi:10.1038/321209a0

8. Song J, Teplova M, Ishibe-Murakami S, Patel DJ (2012) Structure-based mechanistic insights into DNMT1-mediated maintenance DNA methylation. Science 335(6069):709–712. doi:10.1126/science.1214453

9. Chedin F (2011) The DNMT3 family of mammalian de novo DNA methyltransferases. Prog Mol Biol Transl Sci 101:255–285. doi:10.1016/b978-0-12-387685-0.00007-x

10. Rhee I, Bachman KE, Park BH, Jair KW, Yen RW, Schuebel KE, Cui H, Feinberg AP, Lengauer C, Kinzler KW, Baylin SB, Vogelstein B (2002) DNMT1 and DNMT3b cooperate to silence genes in human cancer cells. Nature 416(6880):552–556. doi:10.1038/416552a

11. Lister R, Pelizzola M, Dowen RH, Hawkins RD, Hon G, Tonti-Filippini J, Nery JR, Lee L, Ye Z, Ngo QM, Edsall L, Antosiewicz-Bourget J, Stewart R, Ruotti V, Millar AH, Thomson JA, Ren B, Ecker JR (2009) Human DNA methylomes at base resolution show widespread epigenomic differences. Nature 462(7271):315–322. doi:10.1038/nature08514

12. Cooper DN, Krawczak M (1989) Cytosine methylation and the fate of CpG dinucleotides in vertebrate genomes. Hum Genet 83(2):181–188

13. Takai D, Jones PA (2002) Comprehensive analysis of CpG islands in human chromosomes 21 and 22. Proc Natl Acad Sci U S A 99(6):3740–3745. doi:10.1073/pnas.052410099

14. Chodavarapu RK, Feng S, Bernatavichute YV, Chen PY, Stroud H, Yu Y, Hetzel JA, Kuo F, Kim J, Cokus SJ, Casero D, Bernal M, Huijser P, Clark AT, Kramer U, Merchant SS, Zhang X, Jacobsen SE, Pellegrini M (2010) Relationship between nucleosome positioning and DNA

methylation. Nature 466(7304):388–392. doi:10.1038/nature09147

15. Di Leva G, Croce CM (2013) miRNA profiling of cancer. Curr Opin Genet Dev 23(1):3–11. doi:10.1016/j.gde.2013.01.004

16. Sun K, Lai EC (2013) Adult-specific functions of animal microRNAs. Nat Rev Genet 14(8):535–548. doi:10.1038/nrg3471

17. Baer C, Claus R, Plass C (2013) Genome-wide epigenetic regulation of miRNAs in cancer. Cancer Res 73(2):473–477. doi:10.1158/0008-5472.can-12-3731

18. Iorio MV, Piovan C, Croce CM (2010) Interplay between microRNAs and the epigenetic machinery: an intricate network. Biochim Biophys Acta 1799(10–12):694–701. doi:10.1016/j.bbagrm.2010.05.005

19. Ulitsky I, Bartel DP (2013) lincRNAs: genomics, evolution, and mechanisms. Cell 154(1):26–46. doi:10.1016/j.cell.2013.06.020

20. Franke A, McGovern DP, Barrett JC, Wang K, Radford-Smith GL, Ahmad T, Lees CW, Balschun T, Lee J, Roberts R, Anderson CA, Bis JC, Bumpstead S, Ellinghaus D, Festen EM, Georges M, Green T, Haritunians T, Jostins L, Latiano A, Mathew CG, Montgomery GW, Prescott NJ, Raychaudhuri S, Rotter JI, Schumm P, Sharma Y, Simms LA, Taylor KD, Whiteman D, Wijmenga C, Baldassano RN, Barclay M, Bayless TM, Brand S, Buning C, Cohen A, Colombel JF, Cottone M, Stronati L, Denson T, De Vos M, D'Inca R, Dubinsky M, Edwards C, Florin T, Franchimont D, Gearry R, Glas J, Van Gossum A, Guthery SL, Halfvarson J, Verspaget HW, Hugot JP, Karban A, Laukens D, Lawrance I, Lemann M, Levine A, Libioulle C, Louis E, Mowat C, Newman W, Panes J, Phillips A, Proctor DD, Regueiro M, Russell R, Rutgeerts P, Sanderson J, Sans M, Seibold F, Steinhart AH, Stokkers PC, Torkvist L, Kullak-Ublick G, Wilson D, Walters T, Targan SR, Brant SR, Rioux JD, D'Amato M, Weersma RK, Kugathasan S, Griffiths AM, Mansfield JC, Vermeire S, Duerr RH, Silverberg MS, Satsangi J, Schreiber S, Cho JH, Annese V, Hakonarson H, Daly MJ, Parkes M (2010) Genome-wide meta-analysis increases to 71 the number of confirmed Crohn's disease susceptibility loci. Nat Genet 42(12):1118–1125. doi:10.1038/ng.717

21. Jostins L, Ripke S, Weersma RK, Duerr RH, McGovern DP, Hui KY, Lee JC, Schumm LP, Sharma Y, Anderson CA, Essers J, Mitrovic M, Ning K, Cleynen I, Theatre E, Spain SL, Raychaudhuri S, Goyette P, Wei Z, Abraham C, Achkar JP, Ahmad T, Amininejad L, Ananthakrishnan AN, Andersen V, Andrews JM, Baidoo L, Balschun T, Bampton PA, Bitton A, Boucher G, Brand S, Buning C, Cohain A, Cichon S, D'Amato M, De Jong D, Devaney KL, Dubinsky M, Edwards C, Ellinghaus D, Ferguson LR, Franchimont D, Fransen K, Gearry R, Georges M, Gieger C, Glas J, Haritunians T, Hart A, Hawkey C, Hedl M, Hu X, Karlsen TH, Kupcinskas L, Kugathasan S, Latiano A, Laukens D, Lawrance IC, Lees CW, Louis E, Mahy G, Mansfield J, Morgan AR, Mowat C, Newman W, Palmieri O, Ponsioen CY, Potocnik U, Prescott NJ, Regueiro M, Rotter JI, Russell RK, Sanderson JD, Sans M, Satsangi J, Schreiber S, Simms LA, Sventoraityte J, Targan SR, Taylor KD, Tremelling M, Verspaget HW, De Vos M, Wijmenga C, Wilson DC, Winkelmann J, Xavier RJ, Zeissig S, Zhang B, Zhang CK, Zhao H, Silverberg MS, Annese V, Hakonarson H, Brant SR, Radford-Smith G, Mathew CG, Rioux JD, Schadt EE, Daly MJ, Franke A, Parkes M, Vermeire S, Barrett JC, Cho JH (2012) Host-microbe interactions have shaped the genetic architecture of inflammatory bowel disease. Nature 491(7422):119–124. doi:10.1038/nature11582

22. Sato F, Shibata D, Harpaz N, Xu Y, Yin J, Mori Y, Wang S, Olaru A, Deacu E, Selaru FM, Kimos MC, Hytiroglou P, Young J, Leggett B, Gazdar AF, Toyooka S, Abraham JM, Meltzer SJ (2002) Aberrant methylation of the HPP1 gene in ulcerative colitis-associated colorectal carcinoma. Cancer Res 62(23):6820–6822

23. Ventham NT, Kennedy NA, Nimmo ER, Satsangi J (2013) Beyond gene discovery in inflammatory bowel disease: the emerging role of epigenetics. Gastroenterology 145(2):293–308. doi:10.1053/j.gastro.2013.05.050

24. Megiorni F, Mora B, Bonamico M, Barbato M, Montuori M, Viola F, Trabace S, Mazzilli MC (2008) HLA-DQ and susceptibility to celiac disease: evidence for gender differences and parent-of-origin effects. Am J Gastroenterol 103(4):997–1003. doi:10.1111/j.1572-0241.2007.01716.x

25. Fernandez-Jimenez N, Castellanos-Rubio A, Plaza-Izurieta L, Irastorza I, Elcoroaristizabal X, Jauregi-Miguel A, Lopez-Euba T, Tutau C, de Pancorbo MM, Vitoria JC, Bilbao JR (2014) Coregulation and modulation of NFkappaB-related genes in celiac disease: uncovered aspects of gut mucosal inflammation. Hum Mol Genet 23(5):1298–1310. doi:10.1093/hmg/ddt520

26. Glauben R, Batra A, Fedke I, Zeitz M, Lehr HA, Leoni F, Mascagni P, Fantuzzi G, Dinarello CA, Siegmund B (2006) Histone

hyperacetylation is associated with amelioration of experimental colitis in mice. J Immunol 176(8):5015–5022

27. Glauben R, Batra A, Stroh T, Erben U, Fedke I, Lehr HA, Leoni F, Mascagni P, Dinarello CA, Zeitz M, Siegmund B (2008) Histone deacetylases: novel targets for prevention of colitis-associated cancer in mice. Gut 57(5):613–622. doi:10.1136/gut.2007.134650

28. Glauben R, Siegmund B (2011) Inhibition of histone deacetylases in inflammatory bowel diseases. Mol Med 17(5–6):426–433. doi:10.2119/molmed.2011.00069

29. Rampertab SD, Forde KA, Green PH (2003) Small bowel neoplasia in coeliac disease. Gut 52(8):1211–1214

30. Swinson CM, Slavin G, Coles EC, Booth CC (1983) Coeliac disease and malignancy. Lancet 1(8316):111–115

31. Bergmann F, Singh S, Michel S, Kahlert C, Schirmacher P, Helmke B, Von Knebel Doeberitz M, Kloor M, Blaker H (2010) Small bowel adenocarcinomas in celiac disease follow the CIM-MSI pathway. Oncol Rep 24(6):1535–1539

32. Diosdado B, Buffart TE, Watkins R, Carvalho B, Ylstra B, Tijssen M, Bolijn AS, Lewis F, Maude K, Verbeke C, Nagtegaal ID, Grabsch H, Mulder CJ, Quirke P, Howdle P, Meijer GA (2010) High-resolution array comparative genomic hybridization in sporadic and celiac disease-related small bowel adenocarcinomas. Clin Cancer Res 16(5):1391–1401. doi:10.1158/1078-0432.ccr-09-1773

33. Capuano M, Iaffaldano L, Tinto N, Montanaro D, Capobianco V, Izzo V, Tucci F, Troncone G, Greco L, Sacchetti L (2011) MicroRNA-449a overexpression, reduced NOTCH1 signals and scarce goblet cells characterize the small intestine of celiac patients. PLoS One 6(12):e29094. doi:10.1371/journal.pone.0029094

34. Noonan EJ, Place RF, Pookot D, Basak S, Whitson JM, Hirata H, Giardina C, Dahiya R (2009) miR-449a targets HDAC-1 and induces growth arrest in prostate cancer. Oncogene 28(14):1714–1724. doi:10.1038/onc.2009.19

35. Gray SG (2011) Targeting Huntington's disease through histone deacetylases. Clin Epigenetics 2(2):257–277. doi:10.1007/s13148-011-0025-7

36. Gray SG (2013) Perspectives on epigenetic-based immune intervention for rheumatic diseases. Arthritis Res Ther 15(2):207. doi:10.1186/ar4167

37. Lu Q (2013) The critical importance of epigenetics in autoimmunity. J Autoimmun 41:1–5. doi:10.1016/j.jaut.2013.01.010

38. Edwards AJ, Pender SL (2011) Histone deacetylase inhibitors and their potential role in inflammatory bowel diseases. Biochem Soc Trans 39(4):1092–1095. doi:10.1042/bst0391092

39. Frommer M, McDonald LE, Millar DS, Collis CM, Watt F, Grigg GW, Molloy PL, Paul CL (1992) A genomic sequencing protocol that yields a positive display of 5-methylcytosine residues in individual DNA strands. Proc Natl Acad Sci U S A 89(5):1827–1831

40. Wang RY, Gehrke CW, Ehrlich M (1980) Comparison of bisulfite modification of 5-methyldeoxycytidine and deoxycytidine residues. Nucleic Acids Res 8(20):4777–4790

41. Murphy TM, Sullivan L, Lane C, O'Connor L, Barrett C, Hollywood D, Lynch T, Lawler M, Perry AS (2011) In silico analysis and DHPLC screening strategy identifies novel apoptotic gene targets of aberrant promoter hypermethylation in prostate cancer. Prostate 71(1):1–17. doi:10.1002/pros.21212

42. Perry AS, Liyanage H, Lawler M, Woodson K (2007) Discovery of DNA hypermethylation using a DHPLC screening strategy. Epigenetics 2(1):43–49

43. Prencipe M, McGoldrick A, Perry AS, O'Grady A, Phelan S, McGrogan B, Fitzpatrick P, Watson JA, Furlong F, Brennan DJ, Lawler M, Kay E, McCann A (2010) MAD2 downregulation in hypoxia is independent of promoter hypermethylation. Cell Cycle 9(14):2856–2865

44. Wojdacz TK, Dobrovic A, Hansen LL (2008) Methylation-sensitive high-resolution melting. Nat Protoc 3(12):1903–1908. doi:10.1038/nprot.2008.191

45. Perry AS, Loftus B, Moroose R, Lynch TH, Hollywood D, Watson RW, Woodson K, Lawler M (2007) In silico mining identifies IGFBP3 as a novel target of methylation in prostate cancer. Br J Cancer 96(10):1587–1594. doi:10.1038/sj.bjc.6603767

46. Einarsdottir E, Koskinen LL, Dukes E, Kainu K, Suomela S, Lappalainen M, Ziberna F, Korponay-Szabo IR, Kurppa K, Kaukinen K, Adany R, Pocsai Z, Szeles G, Farkkila M, Turunen U, Halme L, Paavola-Sakki P, Not T, Vatta S, Ventura A, Lofberg R, Torkvist L, Bresso F, Halfvarson J, Maki M, Kontula K, Saarialho-Kere U, Kere J, D'Amato M, Saavalainen P (2009) IL23R in the Swedish, Finnish, Hungarian and Italian populations: association with IBD and psoriasis, and linkage to celiac disease. BMC Med Genet 10:8. doi:10.1186/1471-2350-10-8

47. Fernandez S, Molina IJ, Romero P, Gonzalez R, Pena J, Sanchez F, Reynoso FR, Perez-Navero JL, Estevez O, Ortega C, Santamaria M (2011) Characterization of gliadin-specific

Th17 cells from the mucosa of celiac disease patients. Am J Gastroenterol 106(3):528–538. doi:10.1038/ajg.2010.465

48. Harris KM, Fasano A, Mann DL (2008) Cutting edge: IL-1 controls the IL-23 response induced by gliadin, the etiologic agent in celiac disease. J Immunol 181(7):4457–4460

49. Eads CA, Danenberg KD, Kawakami K, Saltz LB, Blake C, Shibata D, Danenberg PV, Laird PW (2000) MethyLight: a high-throughput assay to measure DNA methylation. Nucleic Acids Res 28(8):E32

50. Yegnasubramanian S, Kowalski J, Gonzalgo ML, Zahurak M, Piantadosi S, Walsh PC, Bova GS, De Marzo AM, Isaacs WB, Nelson WG (2004) Hypermethylation of CpG islands in primary and metastatic human prostate cancer. Cancer Res 64(6):1975–1986

51. Baird AM, Dockry E, Daly A, Stack E, Doherty DG, O'Byrne KJ, Gray SG (2013) IL-23R is epigenetically regulated and modulated by chemotherapy in non-small cell lung cancer. Front Oncol 3:162. doi:10.3389/fonc.2013.00162

Chapter 14

Candidate Gene Knockdown in Celiac Disease

Ben Molloy, Michael Freeley, Aideen Long, and Ross McManus

Abstract

RNA interference (RNAi) is a powerful genetic tool that has created new opportunities in cell biology by allowing the specific modulation of gene expression under controlled conditions. Knockdown of genes associated with disease can provide valuable information pertaining to their function and potentially their role in the disease etiology. In the context of celiac disease, it allows us to examine closely the cellular changes that occur when the expression levels of genes of interest are reduced. Utilizing informative assays that demonstrate changes in cell behavior or other measurable endpoints such as cytokine production or migratory phenotypes can further our understanding of the pathogenic mechanisms in this prevalent auto-immune disorder. This chapter outlines protocols for examining the effects of RNAi on candidate genes and subsequent changes to migratory phenotype, transmigration, and adhesion.

Key words siRNA knockdown, Celiac disease, Migration, Adhesion, T cells

1 Introduction

The discovery of RNA interference has made a significant impact on both basic and applied research. Fire and Mello published their seminal work on RNA interference in *C. elegans* detailing how dsRNA is responsible for posttranscriptional silencing [1]. In 1993, Ambros et al. revealed through cloning that a short noncoding RNA, *Lin-4*, bound the 3′ UTR of its target, *lin-14*, in *C. elegans* [2, 3]. These two pathways of RNA-mediated regulation converge and make use of the same cellular machinery [4]. Researchers can also make use of this cellular machinery to alter specific gene expression directly or indirectly. siRNA can either be transfected directly into the cytoplasm of a cell [5, 6], or alternatively as a vector or plasmid expressing a short hairpin RNA (shRNA) which is subsequently processed into an siRNA [7, 8]. Although siRNAs can be designed and synthesized within labs, the process is normally outsourced to a specialist company.

Celiac disease (CD) is a chronic small intestine immune-mediated enteropathy induced by dietary gluten in genetically

Anthony W. Ryan (ed.), *Celiac Disease: Methods and Protocols*, Methods in Molecular Biology, vol. 1326,
DOI 10.1007/978-1-4939-2839-2_14, © Springer Science+Business Media New York 2015

predisposed individuals [9]. Our current understanding of the pathogenesis of CD has stemmed from a number of landmark findings. Perhaps the most significant finding is the link between the HLA-DQ2 and DQ8 molecules as the major hereditary component of CD and their role in the recognition and initiation of an adaptive immune response against gluten peptides [10]. However, while this step is necessary for development of celiac disease, it is not sufficient. The contribution and identities of other genes/factors are becoming better characterized. GWAS and Immunochip studies have located many non-HLA regions associated with CD. To date, 40 regions containing 64 candidate genes have been identified [11]. These regions generally correspond to LD blocks, which depending on their size may contain anything from gene sections to several candidate genes. The SNPs identified as being disease associated are known as tag SNPs based on their selection as mapping landmarks rather than their potential for functional relevance in disease, and as such are generally not believed to be causal variants (this may not be the case with high-density arrays such as the Immunochip). Although the causal variants have yet to be found, 53 % of CD-associated SNPs correlate with differences in expression levels of at least one physically close gene; these differences are known as *cis* expression quantitative trait loci (*cis* eQTL) SNPs. The number of these cis-eQTL SNPs seen among CD-associated SNPs is much larger than would be expected by chance [12]. This implies that some of the identified risk SNPs influence susceptibility to CD through altered gene expression. With this in mind, knockdown of candidate genes could aid in refining the search for causal variants and in addition provide useful information regarding possible mechanisms of pathogenesis.

When silencing a candidate gene through siRNA, assay choice is extremely important. What output do I want from my knockdown study? What role could my candidate gene be performing in the pathogenesis of CD? The set of non-HLA genes associated with CD appears to be enriched for genes associated with chemokine receptor activity, cytokine binding, T cell activation, and lymphocyte differentiation among others [12]. Villous atrophy, the physical manifestation of CD, is a direct result of damage mediated by cytotoxic IELs expressing activating NK cell receptors that recognize stress and inflammation-induced ligands on intestinal epithelial cells [13]. The migration of IELs into the lamina propria of the gut is therefore a crucial step in the pathogenesis of CD. As a number of candidate genes are associated with chemotaxis, assays examining the effect of knockdown on migration and migratory phenotypes would be informative in the context of CD.

This chapter outlines the relevant protocols for examining the effects knockdown of a candidate gene has on migratory phenotype, transmigration, and adhesion.

2 Materials

Prepare and store all reagents at room temperature unless otherwise stated.

2.1 Extraction and Expansion of Peripheral Blood T Lymphocytes

1. 50 ml Skirted tubes (Sarstedt).
2. 250 ml Containers (Sarstedt).
3. 10 ml Serological Pipettes (Sarstedt).
4. 3.5 ml Transfer Pipettes (Sarstedt).
5. T175 Tissue Culture Flasks, Vented Caps (Sarstedt).
6. Distilled Water (Gibco, Life Technologies/Bio-Sciences).
7. 1× PBS (Gibco, Life Technologies/Bio-Sciences).
8. FBS (Gibco, Life Technologies/Bio-Sciences).
9. Trypan Blue Stain Solution 0.4 % (Gibco, Life Technologies/Bio-Sciences).
10. Red Blood Cell Lysis Buffer (Miltenyi).
11. RPMI-1640 (Sigma).
12. Lymphoprep (Axis-Shield, Norway).
13. PHA (Sigma Aldrich).
14. Recombinant Human IL-2 (Peprotech).

2.2 siRNA Knockdown

1. Tissue Culture 6-well Flat Bottom cell+ (Sarstedt).
2. Cuvettes.
3. 1.5 ml Eppendorfs (Sarstedt).
4. ON-TARGETplus siRNA, 5 nmol (Thermo/Fisher Scientific).
5. ON-TARGETplus Non-Targeting Pool, 5 nmol (Thermo/Fisher Scientific).
6. Mirus Bio Ingenio Electroporation Solution (Mirus/Medical Supply Company).
7. Nucleofactor I, Amaxa Biosystems (Amaxa).

2.3 Transwell, Adhesion, and Polarity Assays

1. Goat Anti-Human IgG (Fc Specific, Sigma Aldrich).
2. Goat Anti-Mouse IgG (Sigma Aldrich).
3. Anti-LFA-1 Antibody (Monosan).
4. Recombinant Human ICAM-1/CD54 (R&D Systems).
5. Recombinant Human SDF-1α (Peprotech).
6. Nunclon 96 well plate (Thermo/Fisher Scientific).
7. 8 % PFA solution—Weigh 8 g of PFA powder (Sigma Aldrich), and add 80 ml of sterile PBS (Gibco) and a few drops of 1 M NaOH (Sigma Aldrich) to a 250 ml glass bottle. Add a magnetic

stirrer and heat to 60 °C in a fume hood. Keep the lid of the bottle loose. Once dissolved, allow the solution to return to room temperature. pH the solution to seven and bring the solution up to 100 ml with PBS. Aliquot and store at –20 °C.

8. Phalloidin-TRITC (Sigma Aldrich).

9. Hoechst 33258 (Molecular Probes).

10. 1× PBS (Gibco, Life Technologies/Bio-Sciences).

11. Tween-20 (Sigma Aldrich).

12. Triton X-100 (Sigma Aldrich).

13. MACS BSA Stock Solution (Miltenyi).

14. 24 well, 12 insert Transwell Plates (Corning, USA).

15. ELISA Plate Sealing Covers (Immunochemistry Technologies).

16. Serum starving medium—RPMI-1640, 0.5 % BSA.

17. Sample buffer Laemmli 2×.

18. Super Signal R. West Dura.

19. Fusion Fx Vilber Lourmat.

2.4 Sodium Dodecyl Sulfate-Polyacrylamide Gel Electrophoresis

1. 30 % Acrylamide (Sigma Aldrich). Store at 4 °C.

2. 10 % w/v ammonium persulfate (APS) (Sigma Aldrich). Store at 4 °C. Make fresh weekly.

3. 1.5 M Tris–HCl pH 8.8—Dissolve 90.825 g of Trizma Base (Sigma Aldrich) in 400 ml of ddH$_2$0. Adjust pH to 8.8 with HCL and bring final volume to 500 ml with ddH$_2$0.

4. 1.0 M Tris–HCl pH 6.8—Dissolve 60.55 g of Trizma Base (Sigma Aldrich) in 400 ml of ddH$_2$0. Adjust pH to 6.8 with HCL and bring final volume to 500 ml with ddH$_2$0.

5. 10 % w/v SDS—Dissolve 10 g of SDS powder (Sigma Aldrich) in 100 ml of ddH$_2$0. Wear a face mask when weighing out SDS powder.

6. TEMED (Sigma Aldrich).

7. Isopropanol (Fisher Scientific).

8. 5×-SDS Running buffer—15.1 g of Trizma (Sigma Aldrich), 94 g of Glycine (Sigma Aldrich), 50 ml of 10 % SDS solution, bring final volume to 1 l with ddH$_2$0.

9. Sample Buffer, Laemmli 2× (Sigma Aldrich).

2.5 Western Blotting

1. Filter paper (VWR).

2. PVDF Membrane (VWR).

3. Methanol (Sigma Aldrich).

4. Scalpel.

5. Western blot transfer buffer—5.8 g of Trizma Base (Sigma Aldrich), 2.9 g of Glycine (Sigma Aldrich), 8 ml of 10 % w/v

sodium dodecyl sulfate (SDS) solution, 200 ml of methanol. Bring final volume to 1 l with ddH$_2$0.

6. 10× Tris-buffered saline (TBS)—24.2 g of Trizma Base (Sigma Aldrich), 80 g of NaCl (Sigma Aldrich), 900 ml of ddH$_2$0. pH the solution to 7.6 with HCl. Bring final volume to 1 l with ddH$_2$0. Dilute to 1× before use.

7. Western blot blocking buffer—5 % w/v Milk Powder (Marvel/TESCO) in western blot wash buffer (TBS-T).

8. Western blot wash buffer—1× TBS with 0.1 % v/v Tween 20 (TBS-T).

9. Super Signal® West Dura Extended Substrate (Thermo/Medical Supply Company).

3   Methods

Perform all steps at room temperature unless otherwise stated.

3.1   Extraction and Expansion of PBTLs

1. Dilute buffy coat (50 ml) in 70 ml of sterile PBS in a disposable, sterile 250 ml container.

2. Pipette 20 ml of Lymphoprep into four sterile skirted 50 ml Falcon tubes. Carefully layer 30 ml of the dilute buffy coat on top of the Lymphoprep in each Falcon (*see* **Note 1**).

3. Centrifuge at 290×g for 20 min. Ensure that the break on the centrifuge is set to zero (*see* **Note 2**).

4. Remove the buffy layer (white blood cell population) from each tube and pool into a new sterile 50 ml tube (*see* **Note 3**). Centrifuge at 450×g for 5 min to pellet the white blood cell population. Remove the supernatant and wash three times with sterile PBS.

5. Incubate the pellet with 5 ml of pre-warmed 1× RBC lysis buffer for 5 min. Centrifuge and resuspend pellet in 10 ml of warm RPMI w 10 % FBS. Add the cell suspension to 40 ml of media and incubate for 1 h at 37 °C in a T175 flask. Ensure that the flask is lying horizontally.

6. Repeat this step for another 1 h with a fresh T175 flask. This step removes monocytes from the white blood cell population (*see* **Note 4**).

7. Count and seed the remaining cells at a concentration between 1×10^6 and 2×10^6 cells/ml. Add PHA to a final concentration of 2 μg/ml. Incubate at 37 °C for 72 h.

8. Wash three times with RPMI and 10 % FBS. Resuspend pellet in desired volume of media with IL-2 at a final concentration of 20 ng/ml. Culture for 5 days. Change media and supplement with IL-2 if necessary (*see* **Note 5**).

| 3.2 *Gene Knockdown in Expanded PBTLs* | After 5 days, the population of cells should be over 95 % T lymphocytes. Prior to counting the population, it is important to decide how many cells you will need. This is dependent on how many genes you wish to knock down in one instance. For simplicity, the following protocol will outline the procedure for knockdown of just one gene. |

1. Wash the cells in RPMI w 10 % FBS once. Resuspend the pellet and count the cells.

2. 5×10^6 cells are required per transfection. Pellet ten million cells and remove all supernatant.

3. Resuspend the pellet in 200 µl of Ingenio® electroporation solution (*see* **Note 6**). Make two 100 µl aliquots.

4. Add target siRNA and control siRNA to the aliquots, respectively, to a final concentration of 1 µM. Flick the tubes gently (*see* **Note 7**).

5. Add the siRNA/cell mixtures to 2 mm electroporation cuvettes. Electroporate using program T-07 on an Amaxa® electroporator (*see* **Note 8**).

6. Pipette the cells into respective wells of a 6-well plate containing 3 ml of RPMI and 10 % FBS. Incubate at 37 °C for 72 h.

7. Between 2 and 4 h post-electroporation, stimulate the cells with 20 ng/ml of IL-2.

3.3 Knockdown Detection

3.3.1 Cell Lysate Preparation

1. Aliquot between 2.5×10^5 and 1×10^6 cells for lysis.

2. Resuspend and wash pellet once in sterile PBS and place on ice.

3. Remove as much supernatant as possible without disturbing the pellet and add 100 µl of ice-cold CERI (*see* **Note 9**) to the pellet, vortex on highest setting for 15 s, and incubate on ice for 10 min.

4. Add 5.5 µl of ice-cold CERII to the tube, vortex for 5 s, and incubate on ice for 1 min.

5. Vortex again for 5 s and centrifuge at $20,000 \times g$ for 5 min in a chilled centrifuge.

6. Immediately remove supernatant, pipette into a fresh tube, and place on ice. This is the cytoplasmic fraction of the cell lysate.

7. Suspend the insoluble pellet in 50 µl of ice-cold NER, vortex for 15 s, and place on ice. Then vortex for 5 s every 10 min for 40 min.

8. Centrifuge at $20,000 \times g$ for 10 min in a chilled centrifuge.

9. Transfer the supernatant to a fresh tube and place on ice. This is the nuclear fraction.

10. Add protease cocktail inhibitors to a final volume of 1×.

11. Add Laemmli buffer in a 1:1 ratio with the supernatant volumes, boil for 5 min, and store at −20 °C (*see* **Note 10**).

Table 1
Constituents of a 10 % acrylamide gel for SDS-PAGE

Component	Volume
dH$_2$O	3.95 ml
30 % Acrylamide	3.35 ml
1.5 M Tris–HCl (pH 8.8)	2.5 ml
10 % w/v SDS	100 µl
10 % w/v APS	100 µl
TEMED	4 µl

Table 2
Constituents of a stacking gel for SDS-PAGE

Component	Volume
dH$_2$O	1.7 ml
30 % Acrylamide	415 µl
1.5 M Tris–HCl (pH 6.8)	315 µl
10 % SDS	25 µl
10 % APS	25 µl
TEMED	2.5 µl

3.3.2 Sodium Dodecyl Sulfate-Polyacrylamide Gel Electrophoresis

1. Rinse electrophoresis plates with ethanol. Attach rubber gaskets to the notched plate and then place the flat plate on top. Fasten on each side with clips.

2. Insert a comb between the plates. Mark 1 cm below the bottom of the comb and remove.

3. Add the components according to Table 1 in sequential order to a 15 ml Falcon to make one 10 % acrylamide resolving gel (*see* **Note 11**).

4. Mix the solution quickly and pipette between the plates up to the 1 cm mark. Overlay the resolving gel with 1 ml of isopropanol (isobutanol can also be used).

5. Allow the gel to set for approximately 45 min.

6. Add the components according to Table 2 in sequential order to a 15 ml Falcon to make a stacking gel.

7. Pour off the isopropanol layer and rinse the top of the gel with a small amount of sterile water.

8. Mix solution quickly and pipette solution over the resolving gel until the level reaches the top of the plate.

9. Place the comb in carefully and leave to set for approximately 45 min.

10. Pour 1× sodium dodecyl sulfate-polyacrylamide gel electrophoresis (SDS-PAGE) running buffer into the electrophoresis rig up to a depth of about 2.5 cm. Place your set gel plate(s) with comb(s) in place into the rig. Fasten the plates with clear plastic holders (*see* **Note 12**).

11. Pour 1× SDS-PAGE running buffer between the plates until it begins to pour down the sides.

12. Remove comb and using a pipette tip, adjust any wells that are not aligned correctly. Flush out wells using a small Pasteur pipette.

13. Load cell lysates (approx. 30 μl) which have been previously boiled for 5 min with Laemmli loading buffer. In addition, load 20 μl of pre-stained broad-range protein ladder that has been boiled under identical conditions.

14. Attach leads and run the gel at 200 V and 25 mA for approximately 1.5 h or until the dye front runs off (*see* **Note 13**).

3.3.3 Western Blotting

1. Remove the gel plate from the rig and carefully separate the plates using a plastic ruler. Cut off the stacking gel (*see* **Note 14**).

2. Soak ten sheets of filter paper (7 cm × 10 cm) in western blot transfer buffer and place on the bottom of an Atto Western Blot Transfer System (*see* **Note 15**).

3. Soak a piece of PVDF membrane (7 cm × 10 cm) in methanol for 30 s. Rinse briefly in transfer buffer and place on top of the soaked sheets of filter paper.

4. Gently remove the gel from between the plates and soak briefly in western blot transfer buffer. Place on top of the PVDF membrane, taking care to exclude any air bubbles.

5. Soak a second set of ten sheets of filter paper (7 cm × 10 cm) in transfer buffer and place on top of the gel to complete the western blot "sandwich" (*see* **Note 16**).

6. Remove any excess transfer buffer from the bottom of the rig.

7. Lower the top of the transfer system, connect the leads, and run at 100 V and 100 mA per gel sandwich for 1 h.

8. Remove and discard the filter paper and gel. Incubate the PVDF membrane in western blot blocking (5 % Marvel in TBST) buffer for 1 h.

9. Wash the PVDF membrane three times for 5 min in each instance with western blot wash buffer.

10. Incubate PVDF membrane with primary antibody (diluted according to the manufacturer's instructions) in western blot blocking buffer overnight at 4 °C with gentle agitation (*see* **Note 17**).

11. Wash the PVDF membrane three times for 5 min in each instance with western blot wash buffer.

12. Incubate PVDF membrane with HRP-labeled secondary antibody (diluted according to the manufacturer's instructions) in western blot blocking buffer for 1–2 h with gentle agitation.

13. Wash the PVDF membrane at least three times for 5 min in each instance with western blot wash buffer.

14. Add 1.5 ml of Super Signal® West Dura—Stable Peroxide Buffer to 1.5 ml of Super Signal West Dura—Luminol/Enhancer Solution in a 15 ml Falcon covered in aluminum foil (*see* **Note 18**).

15. Place the PVDF membrane on a plastic membrane in a chemiluminesence detector (Fusion FX Vilber Lourmat). Pipette the development solution over the PVDF membrane, close the chamber, and leave in darkness for 5 min.

16. Expose the membrane either automatically or manually using the onscreen tabs.

17. Use white light to detect the pre-stained ladder.

3.4 Gene Knockdown in Hut78 Cell Line

While it is more physiologically relevant to conduct knockdown experiments on primary T cell populations, it may not be feasible in some circumstances. To that end, T cell lines can still provide valuable information pertaining to gene function. The protocol for gene knockdown in Hut78 cells differs from the knockdown in expanded peripheral blood T lymphocytes (PBTLs).

3.4.1 Culture of Hut78 Cell Line

1. Cells are cultured at 37 °C and 5 % CO_2 in a humidified atmosphere in either T25 or T75 flasks with ventilated caps.

2. Cells are seeded and cultured at 0.5×10^5 cells/ml.

3. Cultures are split between two and three times a week, dependent on cell growth (*see* **Note 19**). The color of the media will indicate when cells need to be split (yellow culture indicates that cells need to be split).

4. When splitting cells, centrifuge cells at $450 \times g$ for 5 min. Remove supernatant and resuspend pellet in fresh RPMI and 10 % FBS.

3.4.2 Gene Knockdown

1. Count the number of Hut78 cells (*see* **Note 19**).

2. Resuspend 2×10^6 cells in 1 ml of cold electroporation media (RPMI with 10 % FBS).

3. Make 100 μl aliquots depending on the number of knockdowns being carried out.

4. Knockdown and control siRNA is added to each aliquot, respectively, to a final concentration of 500 nM.

5. The cell/siRNA complex is added to a 2 mm gap cuvette and allowed to sit for 2–3 min.

6. The cuvettes are electroporated at 130 mV for 12 ms. The cells are then added to 2 ml of pre-warmed RPMI w 10 % FBS for 24 h at 37 °C.

7. Repeat the electroporation procedure after 24 h and incubate at 37 °C for a further 48 h (*see* **Note 20**).

3.5 Assaying Genetic Knockdown Effect

A number of parameters can be analyzed to determine the effect of knocking down expression of a target gene. As celiac disease is associated with an abnormally high number of immune cells in the lining of the gut, migration is an appropriate parameter to investigate. A number of different assays can be conducted to examine how specific genetic knockdown can alter the ability to adhere, migrate, or adopt a migratory phenotype. Three different assays will be outlined that analyze differences in cell shape, migratory potential, and adhesion.

3.5.1 Cell Polarity Assay

This assay is based on cell morphology in response to a chemokine stimulus.

1. Coat a Nunc 96 well plate with 5 μg/ml goat anti-mouse IgG diluted in sterile PBS. Store at 4 °C overnight.

2. Remove unbound solution carefully and add 1:500 dilution of anti-LFA-1 in sterile PBS or a 1 μg/ml solution of ICAM-1. Incubate for 1–2 h at 37 °C. In addition coat nonmigratory control wells with poly-l-lysine and incubate for 1–2 h at 37 °C.

3. Carefully remove unbound solution and add 100 μl of Hut78 cells (1×10^5 cells/ml) or expanded PBTLs (3×10^5 cells/ml) to the wells.

4. Incubate cells for 1–4 h at 37 °C.

5. Add 100 μl of 8 % PFA to each well and incubate for 15 min at 37 °C.

6. Remove all liquid from wells and wash wells once with PBS and 0.01 % Tween.

7. Permeabilize the cells with PBS and 0.3 % Triton for 5 min.

8. Block the wells with PBS and 3 % BSA for 30 min.

9. Wash the wells three times with PBS and 0.01 % Tween.

10. Stain cells with phallodin-TRITC (1:1000 dilution in PBS) and Hoechst 33258 (1:2000 dilution in PBS) for 30 min in the dark at room temperature.

11. Wash three times with PBS and 0.01 % Tween and store plate(s) at 4 °C until ready for analysis.

12. Image the plates under 20× magnification using an IN Cell Analyzer High Content Analysis (HCA) Imaging System (GE Healthcare, Little Chalfont, UK) or similar system for cell population analysis.

13. Analyze images using accompanying morphology analysis software. Parameters that can be analyzed include cell area, nuclear displacement, cell gyration, and form factor (*see* **Note 21**).

3.5.2 Transwell Migration Assay

This assay mimics migration of cells across a barrier. The system consists of a lower and upper chamber separated by an artificial membrane. This membrane can be coated with an integrin, for example ICAM-1. The lower chamber contains a chemokine in solution. When cells are placed in the upper chamber they make contact with the coated integrin and move across the membrane in response to a chemokine stimulus. This system can be used as an informative in vitro model of trans-endothelial migration, among others.

1. Coat insert membrane carefully with 100 µl of a 1:500 dilution of human IgG and store at 4 °C overnight.

2. Remove unbound solution and add 100 µl of a 1 µg/ml solution of ICAM-1 in sterile PBS and incubate for 2 h at 37 °C.

3. Meanwhile, count the number of cells in each sample and resuspend at a concentration of 2.5×10^5 cells/ml in serum-starved media and incubate at 37 °C for 2 h (*see* **Note 22**).

4. Add 600 µl of RPMI and 0.5 % BSA to the lower chambers of the Transwell plate and add the desired amount of chemokine, in this case SDF-1α.

5. Remove unbound ICAM-1 solution from the well membranes and add 100 µl of serum-starved PBTLs on top of the membrane.

6. Incubate for 2 h at 37 °C.

7. Coat a plate with poly-l-lysine (80 µl) per well and incubate for 1 h at 37 °C (*see* **Note 23**).

8. Remove unbound solution from the wells, add the media from the lower chambers to the poly-l-lysine-coated wells, and incubate for a further 1 h at 37 °C.

9. After 1 h remove half the volume of media from the wells and add an equal volume of 8 % PFA containing 2 µg/ml of cell-permeable Hoechst nuclear dye. Incubate overnight at 37 °C.

10. Remove unbound solution from wells and add 300 µl of sterile PBS to each well. Store at 4 °C until ready to analyze (*see* **Note 24**).

3.5.3 Adhesion Assay

This assay quantifies the percentage of cells in a sample that adhere to an integrin-coated well.

1. Coat wells with 100 μl of a 1:500 dilution of human IgG in sterile PBS and store at 4 °C overnight.

2. Remove unbound solution, add 100 μl of a 1 μg/ml solution of ICAM-1 in sterile PBS to the wells, and incubate for 2 h at 37 °C.

3. Meanwhile count the number of cells in each sample, resuspend at a concentration of 5×10^5 cells/ml in serum-starved media, and incubate for 2 h.

4. Block wells with PBS and 1 % BSA solution for 30 min at 37 °C.

5. Make 100 μl aliquots and add desired amount of SDF-1 α chemokine to each one.

6. Incubate for 30 min at 37 °C.

7. Fill all wells with PBS and seal with a clear adhesive plate cover. Invert the plate and centrifuge at $10 \times g$ for 15 s.

8. Remove the liquid and fix cells with 4 % PFA with 2 μg/ml cell-permeable Hoechst nuclear dye. Incubate overnight at 37 °C.

9. Remove unbound solution from wells and add 300 μl of sterile PBS. Store at 4 °C until ready to analyze.

4   Notes

1. Pipette the dilute blood very slowly at an angle that is close to horizontal. A Pasteur pipette can be used if preferred. It is vital that the layer between the dilute blood and Lymphoprep is not disrupted.

2. The acceleration of the centrifuge can be maintained at the maximum value but the break must be turned off. If the break is left on the rapid deceleration will disrupt the layers in the solution making extraction of the white blood cell population difficult.

3. Using a Pasteur pipette for this step is ideal. Bring the end of the tip towards the "fluffy" layer in the tube. In a simultaneous motion release pressure on the head of the pipette and swirl. Once again, maintaining the integrity of the layers is important. However, disrupting through the red blood cell layer is almost unavoidable but should be minimized.

4. This step can be repeated once more if preferred. Also, the population can be split between two T175 flasks. The number of repeats becomes less important if your aim is to use a column to extract specific T cell populations. This protocol is optimized towards acquiring an expanded PBTL population.

5. As we are dealing with primary T cell populations and not cell lines the differences between individuals can be significant.

Some samples grow aggressively in the presence of IL-2. If the media turns yellow after 48 h of culturing, changing the media would be advisable to maintain your cellular population in a physiologically stable state.

6. The cells must not stay in the Ingenio® solution for longer than 15 min or their integrity will become compromised. For this reason it is important to have everything you need for the experiment prepared prior to resuspension of cells in this solution. The time also limits the number of transfections you can perform in one sitting. I personally have found that I can knock down three target genes and one control comfortably in 12 min. I would not recommend having anymore than six reactions for this time window. If necessary, make a number of cell aliquots and carry out the experiment a number of times.

7. It is advisable to add siRNA to the cells in an Eppendorf where the mixture can be easily allowed to assimilate prior to transfer to a cuvette. I would not recommend adding siRNA directly into the cuvette.

8. The electroporation is successful if a small white frothy head appears on top of the solution in the cuvette post-electroporation.

9. The volumes of reagents CERI and CERII and NER are dependent on packed cell volume. The volumes given are for 1×10^6 cells.

10. Storage at $-20\ ^\circ$C is necessary if you are not going to run the samples in the immediate future. Samples can be stored at 4 $^\circ$C provided that they are analyzed within 5 days.

11. The order in which the components are added is important. Ammonium persulfate (APS) should be made fresh each time you perform a western blot. Only add TEMED when you are ready to pour the resolving gel as the gel will begin to solidify on addition. The volumes specified are enough to make just one gel.

12. I add some running buffer initially so that when placing the plate in I can manipulate it easily to release any air bubbles at the bottom of the gel. Air bubbles will interfere with transmission of charge through the gel. In addition when running just one gel, a second set of plates must still be used. The gel plate should have the lower notched side facing the center of the rig. The empty plate should have the notched side facing away from the center. When using two gels the notched sides should face each other.

13. These settings are for one gel. For two gels, the values are 250 V and 50 mA.

14. Using a ruler to separate the gel plates is safer than using a scalpel. In addition using a scalpel can damage the plates.

15. Although you can use just five sheets of filter paper rather than ten, I have found that my transfer is better when I use more filter paper rather than less. Using just five is perfectly acceptable though.

16. Pouring a small amount of transfer buffer over the sandwich and then smoothing over the top with the sides of your hands moving away from the center of the sandwich ensure that all layers are soaked and air bubble free.

17. Its is advisable to make up primary antibodies in BSA rather than Marvel as they can be reused several times as long as they are kept at 4 °C. Secondary antibodies can be made up in marvel or BSA but should only be used once.

18. These solutions can be made rather than purchased as a kit. The luminol solution is very sensitive to light. In addition western blots can be developed using film.

19. Hut78 cells do not grow as aggressively as other T cell lines. They also like to be maintained at lower cell densities. They should be seeded at 0.5×10^5 cells/ml. In addition, Hut78 cells tend to clump together. Prior to counting cells, the suspension should be pipetted up and down a number of times to enable accurate cell counts. This is important in knock down experiments to get an accurate cell/siRNA ratio.

20. Knocking down gene expression in the same cells twice has been shown to increase knockdown efficiency in certain cases.

21. These are just some parameters that can be analyzed which indicate the degree to which a cell has adopted a migratory phenotype. Cell area and nuclear displacement are easily understood variables. Cell gyration is a measure of "cell spreading" and form factor indicates the "roundness" of a cell. A form factor of 1 implies that the cell is perfectly round and thus in a nonmigratory state.

22. As the expanded PBTL population has been in the presence of IL-2, it is important to culture the cells in serum-starved media (RPMI, 0.5 % BSA) to switch off all IL-2-associated pathways.

23. It is not possible to fix and analyze the cells in the lower chamber of the Transwell plate. Thus the cells must be transferred to the wells of a standard Nunc 96 well plate. Ensure that for each lower chamber, three wells in the poly-l-lysine plate are reserved. It is not advisable to transfer anymore than 300 µl to a single well.

24. The InCell Imaging system or a similar system is used again to image the cells in each well. The number of cells in each well is then calculated using appropriate software. In the case of Transwell migration assays, the number of cells that pass through from the upper to the lower chamber is normalized

against a loading control. The normalized values for knockdowns are then compared against control values to ascertain whether the effect of the knockdown has resulted in more/less cells migrating through the membrane.

Acknowledgement

The authors acknowledge funding from Science Foundation Ireland grant 09/IN.1/B2640 to R.M.M.

References

1. Fire A, Xu S, Montgomery MK, Kostas SA, Driver SE, Mello CC (1998) Potent and specific genetic interference by double-stranded RNA in Caenorhabditis elegans. Nature 391(6669):806–811

2. Lee RC, Feinbaum RL, Ambros V (1993) The C. elegans heterochronic gene lin-4 encodes small RNAs with antisense complementarity to lin-14. Cell 75(5):843–854

3. Wightman B, Ha I, Ruvkun G (1993) Posttranscriptional regulation of the heterochronic gene lin-14 by lin-4 mediates temporal pattern formation in *C. elegans.* Cell 75(5):855–862

4. Shan GE (2010) RNA interference as a gene knockdown technique. Int J Biochem Cell Biol 42:1243–1251

5. Elbashir SM, Lendeckel W, Tuschl T (2001) RNA interference is mediated by 21- and 22-nucleotide RNAs. Genes Dev 15(2):188–200

6. Elbashir SM, Harborth J, Lendeckel W, Yalcin A, Weber K, Tuschl T (2001) Duplexes of 21-nucleotide RNAs mediate RNA interference in cultured mammalian cells. Nature 411(6836):494–498

7. Abbas-Terki T, Blanco-Bose W, Déglon N, Pralong W, Aebischer P (2002) Lentiviral-mediated RNA interference. Hum Gene Ther 13(18):2197–2201

8. Kunath T, Gish G, Lickert H, Jones N, Pawson T, Rossant J (2003) Transgenic RNA interference in ES cell-derived embryos recapitulates a genetic null phenotype. Nat Biotechnol 21(5):559–561

9. Ludvigsson JF, Leffler DA, Bai JC, Biagi F, Fasano A, Green PH, Hadjivassiliou M, Kaukinen K, Kelly CP, Leonard JN et al (2012) The Oslo definitions for coeliac disease and related terms. Gut 62:43–52, Published online 2012

10. Sollid LM (2002) Celiac disease: dissecting a complex inflammatory disorder. Nat Rev Immunol 2:647–655

11. Dubois PC, Trynka G, Franke L, Hunt KA, Romanos J et al (2010) Multiple common variants for celiac disease influencing immune gene expression. Nat Genet 42:395–402

12. Abadie V, Sollid LM, Barreiro LB, Jabri B (2011) Integration of genetic and immunological insights into a model of celiac disease pathogenesis. Annu Rev Immunol 29:493–525

13. Green PH, Jabri B (2003) Celiac disease. Lancet 362:383–391

Part III

Bioinformatics

Chapter 15

Perl One-Liners: Bridging the Gap Between Large Data Sets and Analysis Tools

Karsten Hokamp

Abstract

Computational analyses of biological data are becoming increasingly powerful, and researchers intending on carrying out their own analyses can often choose from a wide array of tools and resources. However, their application might be obstructed by the wide variety of different data formats that are in use, from standard, commonly used formats to output files from high-throughput analysis platforms. The latter are often too large to be opened, viewed, or edited by standard programs, potentially leading to a bottleneck in the analysis. Perl one-liners provide a simple solution to quickly reformat, filter, and merge data sets in preparation for downstream analyses. This chapter presents example code that can be easily adjusted to meet individual requirements. An online version is available at http://bioinf.gen.tcd.ie/pol.

Key words Bioinformatics, Perl, Programming, One-liners, Data merging, Data formatting

1 Introduction

Computational analyses have become widely used by wet-lab scientists and are applied to many different kinds of biological data. For example, most researchers are nowadays familiar with sequence similarity searches and multiple sequence alignments, made popular and accessible through programs such as BLAST [1] and ClustalW [2]. However, there still exist gaps between molecular biology and bioinformatics that prevent smooth data transitions between these two fields. Most of the problems present themselves in the form of incompatible data formats and the need to integrate output from multiple resources. Various kinds of biological measurements, such as sequence data, expression intensities, and structures, can be stored in a multitude of formats, which are not necessarily understood by the program chosen for the analysis. In recently evolved and popular areas, such as prediction of microRNA targets or analysis of RNA-seq data, many different programs exist that one might like to run in parallel for comparison. This can further exacerbate the complexity of input formats. Also, data analyses often require

Anthony W. Ryan (ed.), *Celiac Disease: Methods and Protocols*, Methods in Molecular Biology, vol. 1326,
DOI 10.1007/978-1-4939-2839-2_15, © Springer Science+Business Media New York 2015

the sequential application of multiple programs, which can require data re-formatting as intermediate steps. Finally, output from different resources might have to be combined into one final document.

Data formatting and integration should not be attempted by hand. This is tedious for large data sets and can be prone to introducing errors, for example when merging two data sets with different numbers of rows in a spreadsheet application. Automating such tasks can be achieved with relatively little effort through command-line tools. The Swiss Army knife amongst these is Perl, a scripting language that has featured prominently amongst biologists and bioinformaticists since its development in the late 1980s, as evident from an expansive collection of bioinformatics-related extensions called BioPerl. One of the various meanings of the name Perl is "Practical Extraction and Report Language," which indicates its usefulness for text manipulation and therefore its suitability for data formatting and integration. The following section explains how Perl can be used to solve some common problems that biologists face when trying to carry out bioinformatics analyses on their data. The main focus lies on one-liners, which are one or more instructions written as a single line, without storing the code in a file. This provides an extremely rapid development cycle, eliminates the need of an editor, and avoids potential problems associated with switching between an editor and the command line. One-liners have an upper size limit determined by the maximum allowable line length, usually 256 kb. But in reality they need to be much shorter to be practical. However, a remarkable set of functionality can be achieved even with a few small snippets of code, as demonstrated in the examples below. This chapter provides beginners with a set of examples that can be easily adjusted to meet personal requirements. The experienced programmer might find some new tricks that can be added to their skill set. For ease of use an online version of this chapter has been provided and can be accessed at http://bioinf.gen.tcd.ie/pol.

2 Materials

The Perl examples listed below can be carried out on any computer that features the Perl interpreter and a UNIX-like command-line application (*see* **Note 1**). Apple's operating system, OS X, and most Linux distributions provide these by default. For Windows, several options exist to obtain Perl but the most suitable for the purpose of running one-liners is Cygwin (http://cygwin.com/), which installs not only Perl but also a Linux-like environment on the PC (*see* **Note 2**).

The command line allows executing Perl code either directly or by calling a script stored in a text file. It is also the place from where locally installed bioinformatics programs without graphical user interface can be executed.

The examples shown below are all one-line commands of varying lengths. For more extensive coding one can write and save Perl scripts with the help of a text editor, which can be something as simple as NotePad on Windows or more elaborate tools, such as Text-Wrangler (http://www.barebones.com/products/textwrangler/) or NotePad++ (http://notepad-plus-plus.org/).

3 Methods

3.1 Basic Command-Line Usage

1. Perl one-liners are run from the command line, which is controlled through the keyboard. Each operating system has its own application to invoke the command line. There are also differences in the functionality offered by the shell, which is the mediator between command-line application and the core of the operating system. The Perl one-liners listed below are intended for a command-line interface to a UNIX-like OS. They function the same on each platform but some particularities about working on the command line need to be explained first.

2. Getting started on Mac.

   The default application for the command line on a Mac is called "Terminal." It can be opened by typing the name into Spotfinder or by double-clicking on its symbol in the Applications/Utilities folder.

   A new window appears showing a text line at the top with the cursor positioned after the prompt, which consists of the computer name, the current directory, the user name, and the dollar sign (*see* Fig. 1).

3. Getting started on Windows.

   After installing Cygwin, the command-line interface (also called Bash Shell) can be started by clicking on the according program symbol either on the desktop or under programs. A new window opens showing the log-in and computer name, followed by the current working directory and a prompt in shape of the dollar symbol, followed by a blinking cursor.

4. Getting started on Linux.

   For Linux there are as many command-line applications as there are window managers. In Ubuntu, which uses Gnome as its window manager, a "Terminal" application is available from the Launcher bar, which opens a command-line window when clicked. It shows the prompt, consisting of the log-in and computer name, the current directory name, and the dollar sign, followed by the cursor. Other distributions provide similar mechanisms.

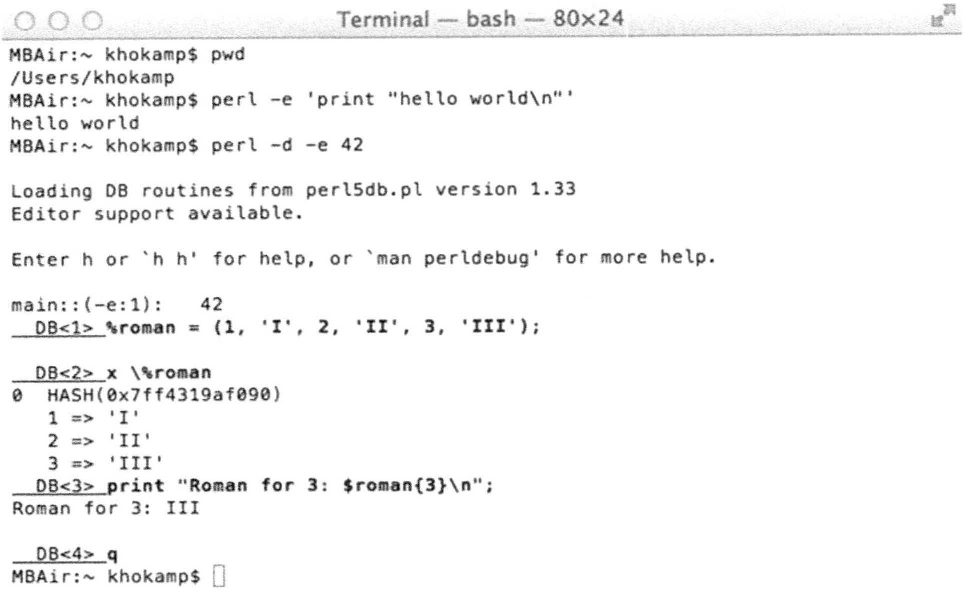

```
 O  O  O                    Terminal — bash — 80×24
MBAir:~ khokamp$ pwd
/Users/khokamp
MBAir:~ khokamp$ perl -e 'print "hello world\n"'
hello world
MBAir:~ khokamp$ perl -d -e 42

Loading DB routines from perl5db.pl version 1.33
Editor support available.

Enter h or `h h' for help, or `man perldebug' for more help.

main::(-e:1):    42
  DB<1> %roman = (1, 'I', 2, 'II', 3, 'III');

  DB<2> x \%roman
0  HASH(0x7ff4319af090)
   1 => 'I'
   2 => 'II'
   3 => 'III'
  DB<3> print "Roman for 3: $roman{3}\n";
Roman for 3: III

  DB<4> q
MBAir:~ khokamp$ []
```

Fig. 1 Screenshot of a Terminal window on a Mac computer. The text in the window shows different commands that were typed (UNIX, Perl one-liner, Perl debugger) and their resulting output

5. The prompt.

   The command line is a place to type instructions for the computer. The start of the line can include a variation of items, such as computer name, log-in name, current working directory, and other user-configurable pieces of information. This is normally followed by a dollar sign ($) or greater-than (>) symbol. After that comes a blinking or highlighted cursor indicating where text will appear. Commands are processed by pressing the "enter" or "return" key (*see* **Note 3**).

6. File system organization.

   The storage space for user accounts is organized through directories anchored to the home directory like roots to a tree. A directory listing from within the home directory normally shows folders like "Desktop," "Documents," and "Downloads." On UNIX-based systems this is achieved through typing "ls -l" followed by hitting return. The output lists files and folders, the latter being indicated by a leading "d" (*see* **Note 4**).

7. Changing directory.

   Check which directory the cursor is currently located in by typing "pwd," which stands for "print working directory." From the home directory change to the Desktop by typing "cd Desktop." From there move to the downloads directory: "cd ../Downloads" (*see* **Note 5**).

8. Paths.

   When working with files, these can be specified through relative or absolute path names. If the current working directory is the Desktop folder, then a file input.txt in the Downloads directory can be specified through "../Downloads/input.txt" (relative path) or "~/Downloads/input.txt" (absolute path) (*see* **Note 6**).

9. Command-line history.

Commands that were typed on the command line are stored in a history file and can be recalled by stepping back using the up-arrow key. The history can also be searched by typing Ctrl-R (control key together with the "R" key) (*see* **Note 7**).

3.2 Perl One-Liners

1. Before delving into more complex Perl code, a few short examples are shown to demonstrate the general use of Perl from the command line. For Perl novices it is advisable to take a look at the Appendix first for a quick introduction to Perl basics.

2. Issue a Perl statement, in this case a greeting message following by a newline, through the -e flag:

```
perl -e 'print "hello world\n"'
```

(*see* **Note 8**).

3. Let Perl add automatically a newline to the output:

```
perl -l -e 'print 2**13'
```

(*see* **Note 9**).

4. Read input from a file and report the number of lines and characters:

```
perl -lne '$i++; $in += length($_); END {
print "$i lines, $in characters"; }' input.txt
```

(*see* **Notes 10** and **11**).

5. Redirect output into a file, in this case 100 random numbers between zero and one:

```
perl -le 'foreach (1..100) { print rand;}'
> random_numbers.txt
```

(*see* **Note 12**).

6. Carry out modifications within one or more files, e.g., by removing all lines starting with a comment (indicated by a hash symbol), but creating a backup file with suffix ".bak" first:

```
perl -p -i.bak -e 's/^#.+//s;' input1.txt
input2.txt input3.txt
```

(*see* **Note 13**).

3.3 Reorganizing Input

1. Reorganizing the column and row order of a text file can be easily accomplished with the -lane flags which rotate through lines and split input at white space or other symbols specified by -F.

2. Swap the order of the second and third columns (indexed with 1 and 2, respectively, in a Perl array) in a tab-delimited file:

```
perl -F"\t" -lane 'print(join "\t",
@F[0,2,1,3..$#F])' input > output
```

(*see* **Note 14**).

3. Bring the last column to the front in a comma-separated file:

```
perl -F, -lane 'print(join ",",
@F[-1,0..$#F-1])' input > output
```

4. Sort in decreasing order by values in the third column of a white-space-delimited file:

```
perl -lane '$s{$F[2]} .= $_; END {foreach
(sort { $b <=> $a } keys %s) { print $s{$_}; }}'
input > output
```

(*see* **Note 15**).

5. Change gene coordinates; that is, make sure that start is always smaller than end coordinate:

```
perl -lane '($F[3],$F[2]) = ($F[2],$F[3])
if ($F[3] < $F[2]); print(join "\t", @F);' input
> output
```

3.4 Filtering Input

1. Perl's text processing features are ideally suited for filtering operations that extract the desired content from a file.

2. Skip empty lines and lines containing only white space:

```
perl -lne 'print if (/\S/)' input > output
```

3. Filter on absolute fold changes and p-values, located in this case in columns 3 and 5 (indexed 2 and 4), respectively:

```
perl -lane 'print if (abs($F[2]) >= 2 and
$F[4] <= 0.05)' input > output
```

4. Extract lines that contain Ensembl gene identifiers:

```
perl -lne 'print if (/ENSG\d+/)' input > output
```

5. Extract lines that contain membrane-related terms, case insensitive:

```
perl -lne 'print if (/membrane/i)' input > output
```

6. Subsampling—extract approximately 1% of random lines from an input file:

```
perl -lne '$i = rand; print if ($i <=
0.01)' input > output
```

(*see* **Note 16**).

3.5 Modifying Input

1. Most of the following examples rely on the -p -i -e flags, which modify text in place. It is advisable to automatically create a backup file first by appending a suffix to the -i flag.

2. Remove all double quotes from files:

```
perl -p -i.bak -e 's/\"//g' input1 input2
input3
```

3. Turn all text into lower case:

```
perl -p -i.bak -e '$_ = lc $_' input*
```

(*see* **Note 17**).

4. Change Windows line ending to UNIX style:

```
perl -p -i.bak -e 's/\cM/\n/g' input*
```

(*see* **Note 18**).

5. Add line numbers to all rows containing text:

```
perl -p -i.bak -e 's/^/sprintf("%-5s",
++$i)/e if (/\S/)' input*
```

(*see* **Note 19**).

6. Trim lines down to the first 42 letters:

```
perl -lne 'print(substr $_, 0, 42)' input
> output
```

7. Reformat the chromosome identifiers in a fasta file, for example when shortening headers such as ">Sc: Oct_2003;chromosome=1" by replacing anything from the greater-than sign to a number with "chr":

```
perl -p -i.bak -e 's/>.+chromosome=(\
d+)/>chr$1/' *fsa
```

8. Round numbers to 1 digit after the decimal point:

```
perl -MRegexp::Common -F"\t" -lane 'foreach
(@F) {$_ = sprintf "%.1f", $_ if (/^$RE{num}{real}$/
and /\./)} print (join "\t", @F)' input > output
```

(*see* **Note 20**).

3.6 Processing Input from One or More Files

1. Perl allows very fine control about how to process and combine input from multiple files. Unfortunately, the code becomes a bit more complex, pushing it to the limit of one-liners.

2. Add columns with minimum and maximum value for each row in a tab-delimited text file:

```
perl -MRegexp::Common -F"\t" -lane '@vals
= (); foreach (@F) { push @vals, $_ if (/^$RE{num}
{real}$/)} @vals = (sort { $a <=> $b } @vals);
print "$_\t$vals[0]\t$vals[-1]"' input > output
```

3. Report frequency of elements in the third column (index 2) of a tab-delimited text file:

```
perl -F"\t" -lane '$freq{$F[2]}++; END
{foreach (sort keys %freq) {print "$_ ->
$freq{$_}";}}' input
```

4. Print the different flags set in a BAM file and how many entries are associated with it. This could be used to check if there is an even number of reads mapping to both strands:

```
      samtools view bam_file | perl -lne '@h =
split "\t", $_; $f{$h[1]}++; END { foreach (sort
keys %f) { print "$_\t$f{$_}";}}'
```

5. Read files with gene IDs and report in decreasing order in how many files each ID is found:

```
      perl -e 'foreach (@ARGV) {open (IN, $_);
while (<IN>) {chomp; $in{$_}++;}} foreach (sort
{ $in{$a} <=> $in{$b} } keys %in) { print "$_ ->
$in{$_}\n";}' file*
```

6. Print the reverse complement of all sequences in a fasta file:

```
      perl -lne 'if (/>/) {$h = $_} else {$in{$h}
.= $_;} END { foreach (sort keys %in) { $s = lc
reverse $in{$_}; $s =~ tr/acgt/tgca/; print "$_\
n$s"}}' input > output
```

7. Extract a subsequence (50 base pairs at position 1000) from a file with a single fasta sequence:

```
      perl -lne 'next if (/^>/); $s .= $_; END
{ print(substr $s, 1000-1, 50) }' input
```

8. Report input lines that differ in their first element between:

```
      perl -e '$f1 = shift; open (IN, $f1); while
(<IN>) {@h = split; $f1{$h[0]}++;} close IN; $f2
= shift; open (IN, $f2); while (<IN>) {@h =
split; $f1{$h[0]}--;} foreach (sort keys %f1) {
print "$_ -> $f1{$_}\n" if ($f1{$_})}' file1 file2
> diff.txt
```

Lines only found in the first file will be printed with a value of 1, and lines only in the second with a value of –1.

9. Use a file of IDs to filter lines from another file and report IDs that were not found:

```
      perl -e '$f = shift; open (IN, $f); while
(<IN>) {@h = split; $f{$h[0]}++;} close IN; $f
= shift; open (IN, $f); while (<IN>) {chomp; @h
= split; if (defined $f{$h[0]}) {print "$_\n";
$f{$h[0]} = 0}} foreach (sort keys %f) {print
STDERR        "not        found:        $_\n"
if ($f{$_})}' filter_file input > output
```

(*see* **Note 21**).

10. Combine two files using IDs from first column as key:

```
      perl -e 'foreach $f (@ARGV) {open (IN,
$f); while (<IN>) {chomp; @h = split /\t/, $_;
$in{$h[0]}{$f} = $_; } close IN; } foreach (sort
keys %in) { print "$in{$_}{$ARGV[0]}\t$in{$_}
{$ARGV[1]}\n";}' input1 input2 > combined
```

(*see* **Note 22**).

11. Print all palindromes of length 15 found in a sequence file using a sliding window approach:

```
perl -lne '$in .= $_; END {foreach (0..
length($in)-15) { $t = substr $in, $_, 15; print
"$i_: $t" if ($t eq reverse($t)) } }' input
```

3.7 Running Programs on Multiple Input Files

1. Perl can be used as a wrapper to apply the same process to multiple input files but naming the output according to the original file.

2. Bowtie mapping of multiple files to output files without the ".fastq" ending:

```
perl -e 'foreach (@ARGV) { ($out = $_) =~
s/\.fastq//; system "bowtie2 index_file \"$_\" >
\"$out.sam\"" 2> \"$out.log\"";}' *.fastq
```

(*see* **Notes 23** and **24**).

3. Transform SAM to BAM files:

```
perl -e 'foreach (@ARGV) {($out = $_) =~
s/\.sam//; system "time samtools view \"$_\" >
\"$out.bam\" 2> $out.err";}' *.sam
```

(*see* **Note 25**).

4. Sort and index BAM files:

```
perl -e 'foreach (@ARGV) {($out = $_) =~
s/\.bam//;  system "time samtools sort \"$_\"
\"$out.sorted"; time samtools index \"$out.
sorted.bam\"";}' *.bam
```

5. Change multiple WIG files into IGB files:

```
perl -e 'foreach (@ARGV) { ($out = $_) =~
s/\.wig/.gr/; system "grep -v variable \"$_\" >
\"$out\"";}' *tpm.wig
```

(*see* **Note 26**).

6. Find all files in the current directory that have not been modified within the last 14 days and print their names together with the date they were last modified:

```
perl -e 'foreach (<*>) { if (-M > 14) {
$age=(stat($_))[9]; print"$_\t".(localtime($age))."\n";}}'
```

(*see* **Note 27**).

3.8 Further Reading

1. This chapter only provides a selection of one-liners—many other applications are possible. An online search for bioinformatics one-liners brings up a variety of blogs that list useful code snippets, not only in Perl but also for other programming languages.

2. An outstanding website for Perl one-liners is the Scriptome at http://sysbio.harvard.edu/csb/resources/computational/

scriptome/. It not only shows useful examples but also provides options for modification of the code and expansion into proper scripts.

3. An introduction to Perl one-liners in general is provided in this chapter from a pre-edition of the Perl Review: http://www.theperlreview.com/articles/one-liners.html.

4. An extensive selection of one-liners is presented and explained in an e-book by Peteris Krumins: http://www.catonmat.net/blog/perl-book/.

5. As with any language, it does not suffice to just read about Perl and one-liners—practice makes perfect!

4 Notes

1. UNIX is an operating system that was developed in the 1970s.

2. The cmd.exe tool or PowerShell in Windows functions quite different to UNIX-style command lines with regard to the use of quotes and therefore renders it unsuitable for the examples listed below. Cygwin provides a command-line interface with UNIX-style functionality.

3. The tilde ("~") is a shortcut for the home directory and may appear within the prompt.

4. Without the -l flag the "ls" command only lists the names of files and folders without detailed information.

5. Two dots beside each other indicate the directory one level up from the current directory. Forward slashes are used as delimiters between folders.

6. Absolute paths normally start from the root "/," and the tilde is a shortcut for the absolute location of the home directory.

7. To increase the limit of commands saved to history, put the following line into the ".bashrc" file in your home directory (assuming you use the Bash Shell): HISTSIZE = 100000.

8. Computers are not tolerant to spelling mistakes. Even the slightest error, e.g., forgetting one of the quotes, would render the statement unusable. If you get stuck in a one-liner that does not seem to terminate, press Ctrl-c (control key together with c) to cancel a running command.

9. With the -l flag Perl strips off-line endings when reading in data and adds newlines to any output that is printed. It can be combined with the -e flag as -le but not the other way around because -e needs to be followed by the actual Perl code.

10. With the -n flag Perl automatically loops through each line of input and stores it in the default variable "$_".

11. The "END" and "BEGIN" subroutines can be used to run code outside the automatic loop over the input lines.

12. The greater-than sign overwrites an existing file—use two greater-than signs (">>") to append to an existing file.

13. With the -p flag Perl loops through the content of a file, reading in line by line into "$_" and processing it with the command specified. Instead of printing the output to the screen, the -i flag leads to editing in place. The optional extension of -i ("bak" in this case) is appended to a backup copy of the original file.

14. The special variable $#F indicates the highest index in the array @F, and the two dots expand the two numbers around them into a list. Change the order of columns by modifying the numbers within square brackets.

15. Sorting is in alphabetical order by default. The "$a" and "$b" variables in the square brackets are placeholders. Reverse their order to sort in decreasing order (higher elements first).

16. By default the rand function returns numbers between 0 and 1.

17. The star is a wild card that can match zero or more characters.

18. Control-M is the line ending that is added when a text file is generated on a Windows computer. This can cause problems on a UNIX-like computer where the line ending is different.

19. The function "sprintf" provides many different formatting options. In this case the numbers are buffered with spaces to a width of five.

20. The -M flag loads an additional module, i.e., Regexp::Common, to avail of extra functionality that is not part of the core set of functions.

21. Printing to STDERR allows separating the normal output (redirected into a file) from warning messages.

22. Some fine-tuning is necessary to deal with cases where an ID is missing from both files. One could, for example, restrict output to only those lines that are present in file 1 by checking for "if (defined $in{$_}{$ARGV[0]})}" before printing.

23. The redirection of the STDERR channel via "2>" saves log messages separately from the SAM output.

24. File names are surrounded by double quotes in case they contain spaces.

25. Bowtie2 output can be saved directly in BAM format using the following pipeline:

```
      bowtie2 index_file input | samtools view
-bS - > output.bam
```

26. The UNIX function "grep" in combination with the "-v" flag returns every line from a file that does not match a given pattern.

27. Besides "-M" Perl provides many other file tests; see perldoc -f -X for a full listing.

Appendix

Perl Basics

Knowledge of a few basic concepts in Perl will lead to a better comprehension of the instructions contained in the one-liners below. It will also allow the user to modify the code and make adjustments according to individual needs. This appendix explains some basic concepts of Perl that are used in Subheading 3.

Perl Variables

A Perl command can contain several elements, such as variables, operators, built-in functions, and key words. Variables provide storage containers for data and come in three types: scalars (e.g., numbers, letters, or strings of characters), arrays (lists of scalars), and hashes (lists of scalars organized into key-value pairs). Very complex constructs are possible but to keep it simple only the most basic aspects are presented here. Each variable is given a name that starts with a symbol ($ for scalars, @ for arrays, and % for hashes) and is followed by alphanumerical characters (a–Z, 0–9), including the underscore. Below are some simple examples of assigning and accessing variables:

```
$attempt = 1;
$date = '11/12/2013';
print "attempt $attempt on $date\n";
@elements = ('CDS', 'mRNA', 'tRNA');
print "First element: $elements[0]\n";
%roman = (1, 'I', 2, 'II', 3, 'III');
print "Roman for 3: $roman{3}\n";
```

An easy way to try out Perl code is the debugger. It can be started by typing "perl -d -e 42" at the command line. This will give a new prompt ("DB<1>") after which Perl statements can be typed for testing. The debugger provides extra functionality, for example examining the content of variables, which can be particularly useful for beginners.

A couple of rules are worth noting from the lines above:

1. Perl statements end with a semicolon.

2. Value assignments happen from right to left; that is, the value to be assigned is on the right-hand side of the equal sign.

3. Text needs to be enclosed in double or single quotes.

4. Variables and special characters (e.g., "\n") or evaluated within double quotes but not single quotes.

5. Lists are enclosed in round brackets with elements separated by comma.

6. List indices start at position zero.

7. To access a single element of an array, the symbol at the start of the variable changes to "$" and the index is specified in square brackets, "[]".

8. To access a specific value in a hash, the symbol at the start of the variable changes to "$" and the lookup key is specified in curly brackets, "{}".

Perl Operators

The next lines of code demonstrate some example use of operators in Perl (some comments are added, starting with "#"):

```
# some standard mathematical operations
# print 3 * (5 + 10) - 2**4;

# processing the content of variables
$total_error = $fp = $fn;

# increase value in $minutes by 30
$minutes += 30;

# increase value in variable $hour by one
$hour++;

# decrease value in variable $remaining by one
$remaining--;

# repeat 'CG' 12 times
$motif = 'CG' x 12;

# the dot concatenates strings and content of
# variables
$chr = 'chr' . $roman{$chr_number};

# two dots create lists by expanding from
# lower to higher border
@hex = (1..9, a..f);
```

Perl Functions

Perl provides many functions that can be applied to the different variable types. A few are listed below and shown with examples:

```
# functions for scalars
$seq_len = length($seq);
$rev_seq = reverse($seq);
$upper_case = uc($seq);
$lower_case = lc($seq);
$codon = substr $seq, 0, 3;

# remove white-space from end of line
chomp $input_line;

# functions for arrays
@array = split //, $string;
```

```
$first_element = shift @array;
$last_element = pop @array;
unshift @array, $first_element;
push @array, $last_element;
@alphabetically_sorted = sort @names;
@numerically_sorted = sort { $a <=> $b } @
values;

# functions for hashes
if (defined $description{$gene}) { print
$description{$gene} } else { print 'not avail-
able'; }
foreach (keys %headers) { print ">$_\
n$headers{$_}\n"; }
```

Loops and Branches

The last two examples introduced the concept of loops and branches. These operate on lists and Boolean expressions, respectively.

A loop is carried out for each element in a list and an if-statement is executed if a test condition is true. Any Perl statement that evaluates to something different to 0 or an empty string is considered true. For tests comparators are available, such as ">," "<," "==," ">=," and "<=" for numbers and "gt," "lt," and "eq" for characters. A common mistake is to use just a single equal sign to check if two variables are equal. In such cases a double equal sign needs to be used to distinguish the comparison from an assignment. See below for examples:

```
# a progress meter for reading in long files:
if ($line % 1000 == 0) { print STDERR " $line
"; }

# collect lines of sequence into one long
# lower-case string:
while (<>) { chomp; $seq .= lc $_; }

# exact motif search
if (substr($seq, $pos, 10) eq $motif) { print
"Motif found at position $pos!\n"; }

# pad number with zeros at the front
$num = '0'.$num until (length($num) >=
$max_len);
```

The line "while (<>) {}" is a special Perl construct that reads line by line from standard input and stores each line in the special variable "$_". A file name specified on the command line would be automatically opened by the shell and fed into the Perl program.

Regular Expressions

One of the most powerful features of Perl is its implementation of regular expressions, which allow matching not only exact text strings but also variable classes of text. Whole books have been written about this topic and a full explanation would go beyond the scope of this chapter. Therefore, only a few basic concepts are explained and demonstrated in the form of examples.

Regular expressions are specified within delimiters ("/" by default) and applied to the content of a variable with the "=~" operator. If a second expression is provided, then the first pattern will be replaced with the second. In addition, modifiers can be used, such as "i" for case-insensitive matches and "g" for global matches, instead of just the first one. Special characters are available to match groups of characters, such as "\w" for any alphanumerical character, "\d" for numbers, and "\s" for white space. The negated class, e.g., not a digit, can be accessed through capital letters, such as "\D," "\W," and "\S." Occurrences can be specified through numbers in curly brackets, e.g., {3} for exactly 3, or {4,10} for 4–10, or {2,} for two or more occurrences of a pattern. Special cases are "+" for one or more matches, "*" for zero or more matches, and "?" for zero or one match. To refer to the matched patterns afterwards, round brackets are used and the special variables $1, $2, ..., depending on how many patterns are specified. The examples below illustrate their usage:

```perl
# search $_ for the word "regulator" (ignoring
# case) and print if found
if (/regulator/i) { print; }

# check for non-numerical input
if ($input =~ /\D/) { warn "Non-numerical
input in '$input'\n"; }

# remove all white space
$input =~ s/\s//g;

# find a pattern that is repeated at least 3
# times and print
if ($input =~ /(CG{3,})/) { print "Found
motif $1!\n"; }

# split a string at tabulators and collect
# the elements in an array
@list = split /\t/, $input;
```

There is plenty of literature available for more information on learning Perl. A good starting point is the online library at perl. org: http://www.perl.org/books/library.html.

References

1. Altschul SF, Gish W, Miller W et al (1990) Basic local alignment search tool. J Mol Biol 215:403–410
2. Thompson JD, Higgins DG, Gibson TJ (1994) CLUSTAL W: improving the sensitivity of progressive multiple sequence alignment through sequence weighting, position-specific gap penalties and weight matrix choice. Nucleic Acids Res 22:4673–4680

Chapter 16

Bioinformatic Analysis of Antigenic Proteins in Celiac Disease

Cathal P. O'Brien

Abstract

Investigation of the chemistry of the gliadin proteins has played an important role in our comprehension of how celiac disease (CoD) develops and progresses as a response to challenge with this immune stimulus. Studies in this area have implicated gut enzymes, tissue transglutaminase-mediated deamidation, and peptide binding affinity for the HLA-DQ2 and DQ8 molecules in disease pathogenesis.

As the number and availability of prolamin sequences increases, the complexity and cost of laboratory analysis will similarly increase. Freely available tools to bioinformatically analyze candidate protein sequences can be employed as a low-cost, high-return preliminary mechanism to focus one's laboratory analyses on the most rewarding sequences. This chapter describes the use of antigen prediction, deamidation prediction, and protease cleavage prediction as may be applied to CoD research.

Key words Celiac disease, Bioinformatics, Antigen, Protease, Tissue transglutaminase

1 Introduction

While the HLA-linked nature of celiac disease has long been established, the characterization of the gliadin and other prolamin peptides as the environmental trigger of the inflammatory process has had an equally important and more long-standing impact on our understanding of the disease. It is fitting therefore that the study by Shan et al. reestablished the cruciality of prolamin protein chemistry to the disease process [1]. The goal of this chapter is to provide an outline of software and services that may be used to investigate and characterize novel, or established, antigenic peptides in the context of CD.

1.1 Prediction of Immunogenic Peptides Using RANKPEP

While prediction of antigen binding to MHC class I molecules (MHC-I) is relatively well developed, techniques that predict peptide binding to MHC class II molecules (MHC-II) are limited by the open-ended binding grooves which are a feature of MHC-II [2]. To date a number of different approaches have been applied to

Anthony W. Ryan (ed.), *Celiac Disease: Methods and Protocols*, Methods in Molecular Biology, vol. 1326,
DOI 10.1007/978-1-4939-2839-2_16, © Springer Science+Business Media New York 2015

the problem of MHCII molecules but in order to simplify this instruction, we will focus on antigen prediction using a well-known server—RANKPEP [3]. RANKPEP uses position-specific scoring matrices to sequentially scan each nonameric subsequence of a target sequence and rank the resultant nonamers based upon their putative ability to bind to the selected MHCII molecule.

1.2 Prediction of Proteolytic Cleavage Using PeptideCutter

As was demonstrated by Shan et al., the resistance of a peptide fragment to enzymatic degradation is a step that, while not sufficient to guarantee antigenicity, does increase the likelihood that a peptide will be seen intact by the immune system [1]. For this reason it may be desirable to examine a protein sequence to investigate potential enzymatic cleavage sites; an abundance of enzymatic cleavage sites may indicate that a region of interest is unlikely to reach the gut mucosa intact.

For this section of the chapter we use PeptideCutter on the ExPASy server which is hosted by the Swiss Institute of Bioinformatics. The PeptideCutter algorithm uses pattern matching to identify sites that may be cleaved by specific enzymes [4]. For the purposes of this analysis, we will focus on the digestive system enzymes pepsin, chymotrypsin, and trypsin; however, a greater variety of enzymes are available to choose from when performing the analysis.

1.3 Prediction of Deamidation of Glutamine Residues Using Regular Expressions

The work of Vader et al. which focused on the identification of deamidation sites in gliadin proteins allowed the authors to identify a number of novel antigens in CoD [5]. Given that deamidated peptides have been shown to be more antigenic to CoD gut-derived T-cells, it is possible that bioinformatic conversion of a peptide to its deamidated equivalent may lead to a more accurate HLA-binding prediction or may act as a standalone indicator of antigenicity. This analysis will use a subset of the deamidation patterns described by Vader et al.

As no online service exists for the bioinformatic identification and conversion of protein sequences to their putative deamidated counterparts we will use regular expressions to identify and convert tTG-targeted glutamine residues to glutamic acid. Regular expressions are a commonly used pattern-matching tool in programming and bioinformatics and are similar to the "Find" and "Replace" functionality present in modern word processors.

While our analysis will use Microsoft® Word, a number of alternative options such as scripting exist for analysis of protein sequences using regular expressions. As scripting and programming methodologies will not be described in this chapter, we outline some of the options for implementing regular expressions using command-line tools or scripting. For those users who will use UNIX® or Linux pipelines and shell scripts, the command-line tools "grep" and "sed" can be used to locate and replace your chosen pattern. Depending on the exact requirements of your analysis, pipelining between programs may be necessary. As with the majority of core UNIX® and Linux programs, support for the commands can be found through the "man pages" for each command. For more complex analyses where multiple sequence manipulations may be necessary, the use of a scripting language such as Python or Perl may be advisable. Both languages have readily available modules for regular expressions and excellent support resources available via the World Wide Web. While Python can be an easier language to learn, each language has its own advantages and each has a community repository of bioinformatics functions available, namely Biopython (www.biopython.org) and BioPerl (www.bioperl.org).

1.4 Conclusion

All of the tools described in this chapter represent good approximations of biological phenomena that have been previously characterized in vitro or in vivo. The exact mechanism by which they are employed would depend on the nature of the study being undertaken. For example, a project focused on identification of novel immunogenic prolamin peptides might use a combination of all three techniques outlined in this chapter, while a more focused project may use one or two of the techniques. Regardless of the order of testing or the scope of bioinformatic analysis employed, it is important that researchers remain mindful of the strengths and weaknesses of the techniques.

2 Materials

The most consistent material requirement for these analyses is an up-to-date computer and web browser with an Internet connection. For each section the details of Web-based or local resources are specified below.

2.1 Prediction
of Immunogenic
Peptides Using
RANKPEP

Web resource: http://imed.med.ucm.es/Tools/rankpep.html.

2.2 Prediction
of Proteolytic Cleavage
Using PeptideCutter

Web resource: http://web.expasy.org/peptide_cutter/.

2.3 Prediction
of Deamidation
of Glutamine Residues
Using Regular
Expressions

A program capable of handling regular expressions will be necessary for this analysis. While a number of programs exist for multiple platforms, Microsoft® Word's "Find and Replace" functionality will be used due to the widespread availability of the program. The downside to using Word is the limited ability to automate the process, a limitation that is common to most graphical user interface-operated programs. Linux users may prefer to use command-line arguments for their regular expressions, and indeed, one should consider such methods if high-volume analyses are likely to be required.

3   Methods

3.1 Prediction of Immunogenic Peptides Using RANKPEP

1. Direct your web browser to the RANKPEP url (http://imed. med.ucm.es/Tools/rankpep.html).

2. From the main page of the RANKPEP server you will first need to select the MHC molecule of interest. For celiac disease research, molecules such as HLA-DQ2 or HLA-DQ8 will be of most interest; however, one can also select from a variety of other MHC class I and class II molecules.

3. Copy the target protein sequence (*see* **Note 1**) using the copy or cut commands from your text editor or word processor. The protein sequence should be in FASTA format (*see* **Note 2**).

4. Ensure that input type is selected as FASTA sequence/s.

5. Select a binding threshold of 5 %. This will return the top 5 % of predicted binding sequences from within the targeted protein sequence(s). The authors of this resource estimate that approximately 80 % of MHCII restricted epitopes are found among the top 5 % of predicted binding peptides.

6. A number of other options may be selected depending on the nature of the analysis being undertaken; however, these options are not immediately relevant for the majority of studies.

7. Once completed the form should resemble the image of the RANKPEP main page (Fig. 1). Pressing the "Send" button should submit the form and generate a list of results.

8. The results section should contain the following sections:

   (a) Consensus: The consensus optimal sequence to bind to your selected MHC molecule.

   (b) Optimal score: The theoretical optimal score for a peptide binding to the selected MHC molecule.

   (c) Binding threshold: A score below which peptides are not predicted to bind to the selected MHC molecule with sufficient affinity to be immunogenic (*see* **Note 3**).

   (d) Table of results: The listing of the top 5 % of immunogenic peptides predicted to bind to the selected MHC molecule based on the figures in the position-specific scoring matrix.

3.2 Prediction of Proteolytic Cleavage Using PeptideCutter

1. Open the PeptideCutter resource by directing your web browser to http://web.expasy.org/peptide_cutter.

2. Paste your query sequence without the FASTA header (the line beginning with the ">" character) into the query window

Rankpep: prediction of binding peptides to Class I and Class II MHC molecules

Description

This server predicts peptide binders to MHCI and MHCII molecules from protein sequence/s or sequence alignments using Position Specific Scoring Matrices (PSSMs). In addition, it predicts those MHCI ligands whose C-terminal end is likely to be the result of proteasomal cleavage. A detailed explanation of the method can be found here.

	SELECT PSSM (Check MHCI or MHCII)
	○ MHC I ⊙ MHC II
PSSM ❷	H2–Db (mouse) [8mer] HLA–DQ1 H2–Db (mouse) [9mer] HLA–DQ1(DQA1*0101xDQB1*0501) H2–Db (mouse) [10mer] HLA–DQ2(DQA1*0501xDQB1*0201) H2–Db (mouse) [11mer] HLA–DQ2(DQA1*0501xDQB1*02) H2–Dd (mouse) [9mer] HLA–DQ5(DQA1*0101xDQB1*0501)
	OR, UPLOAD YOUR PSSM ❷ [Choose File] no file selected
INPUT ❷	TYPE: ⊙ FASTA sequence/s ❷ ○ CLUSTALW multiple sequence alignment ❷ Replace example with your query >tr|R4VEK6|R4VEK6_WHEAT Gliadin OS=Triticum aestivum GN=gli PE=4 SV=1 MKTFLILALLAIVATTATTAVRVPVPQLQPQHPSQQQPQEQVPLVQQQQFLGQQQPFPPQ QPYPQPQPFPSQQPYLQLQPFPQPQLPYSQPQPFRPQQPYPQPQPQYSQPQQPISQRQQQ QQQQQQQQQQQILQQILQQQLIPCMDVVLQQHNIAHGRSQVLQQSTYQLLQELCCQHLWQI OR, UPLOAD SEQUENCES ❷ [Choose File] no file selected
BINDING THRESHOLD ❷	⊙ PERCENTAGE: [2 % ⬍] ○ TOP NUMBER: [990 ⬍]
PROTEASOME CLEAVAGE ❷	FILTER: [OFF ⬍] LMPC ❷: [One ⬍] If Filter is ON only peptides predicted to be cleaved are shown

ADVANCED OPTONS	
RESTRICT RESULTS BY MW ❷ **Lower Limit for Molecular Weight** [0.00] **Upper Limit for Molecular Weight** [9999.00]	**VARIABILITY MASKING** ❷ **Select Variability Threshold** ❷ [1] Value must range between 0.0 and 4.3

[Send] [Clear Form]

Fig. 1 Screenshot of the RANKPEP server home page

(*see* Fig. 2, *see* **Note 4**). A UniProt ID can also be entered into this window if that is preferred.

3. Select the option to use "only the following selection of enzymes and chemicals."

4. Choose the enzymes that are required for analysis (*see* **Note 5**).

5. Select the "Perform" button to commence the analysis.

6. The software will return a report detailing the enzymes that have been included in the analysis, the potential cleavage sites, and the enzymes that are likely to cleave the protein at each site.

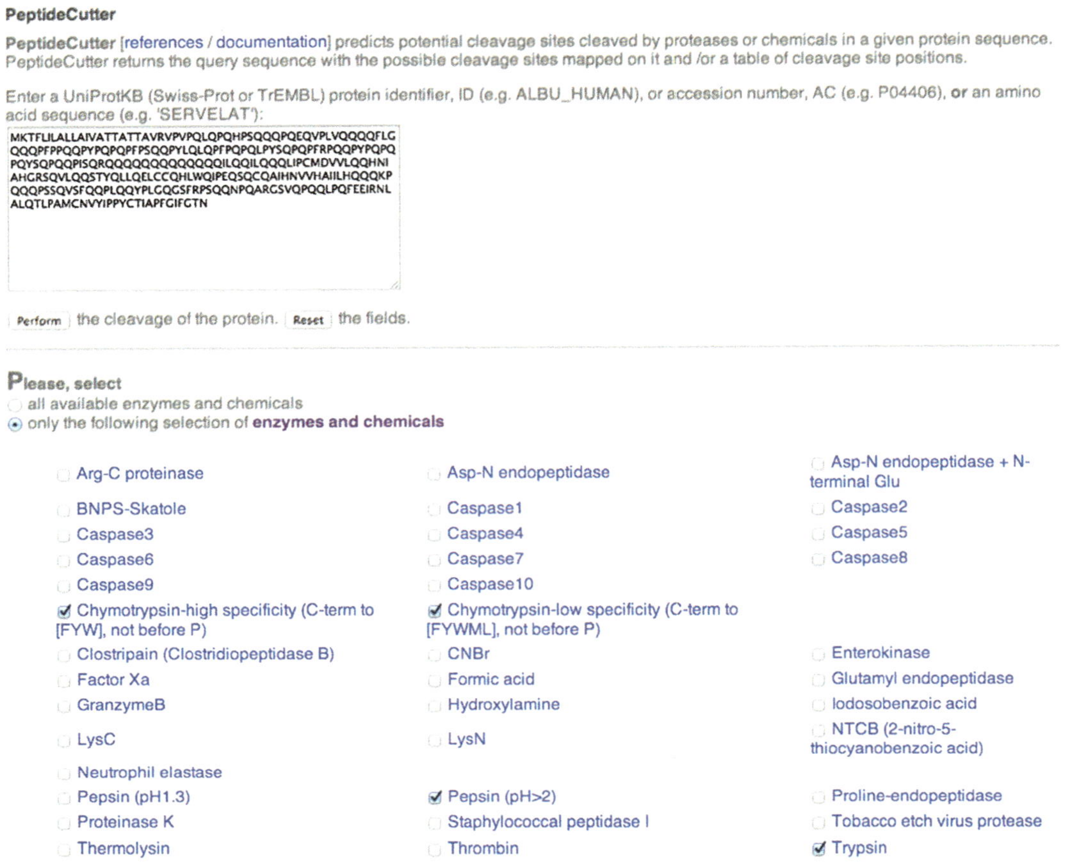

PeptideCutter

PeptideCutter [references / documentation] predicts potential cleavage sites cleaved by proteases or chemicals in a given protein sequence. PeptideCutter returns the query sequence with the possible cleavage sites mapped on it and /or a table of cleavage site positions.

Enter a UniProtKB (Swiss-Prot or TrEMBL) protein identifier, ID (e.g. ALBU_HUMAN), or accession number, AC (e.g. P04406), **or** an amino acid sequence (e.g. 'SERVELAT'):

```
MKTFLILALLAIVATTATTAVRVPVPQLQPQHPSQQQPQEQVPLVQQQQFLG
QQQPFPPQQPYPQPQPFPSQQPYLQLQPFPQPQLPYSQPQPFRPQQPYPQPQ
PQYSQPQQPISQRQQQQQQQQQQQQQILQQILQQQLIPCMDVVLQQHNI
AHGRSQVLQQSTYQLLQELCCQHLWQIPEQSQCQAIHNVVHAIILHQQQKP
QQQPSSQVSFQQPLQQYPLGQCSFRPSQQNPQARGSVQPQQLPQFEEIRNL
ALQTLPAMCNVYIPPYCTIAPFGIFGTN
```

  Perform   the cleavage of the protein.   Reset   the fields.

Please, select
○ all available enzymes and chemicals
◉ only the following selection of **enzymes and chemicals**

○ Arg-C proteinase	○ Asp-N endopeptidase	○ Asp-N endopeptidase + N-terminal Glu
○ BNPS-Skatole	○ Caspase1	○ Caspase2
○ Caspase3	○ Caspase4	○ Caspase5
○ Caspase6	○ Caspase7	○ Caspase8
○ Caspase9	○ Caspase10	
☑ Chymotrypsin-high specificity (C-term to [FYW], not before P)	☑ Chymotrypsin-low specificity (C-term to [FYWML], not before P)	
○ Clostripain (Clostridiopeptidase B)	○ CNBr	○ Enterokinase
○ Factor Xa	○ Formic acid	○ Glutamyl endopeptidase
○ GranzymeB	○ Hydroxylamine	○ Iodosobenzoic acid
○ LysC	○ LysN	○ NTCB (2-nitro-5-thiocyanobenzoic acid)
○ Neutrophil elastase		
○ Pepsin (pH1.3)	☑ Pepsin (pH>2)	○ Proline-endopeptidase
○ Proteinase K	○ Staphylococcal peptidase I	○ Tobacco etch virus protease
○ Thermolysin	○ Thrombin	☑ Trypsin

Fig. 2 Screenshot of the PeptideCutter service page at ExPASy

Table 1
Characterized deamidation patterns in gliadin peptides from Vader et al.'s paper [5]

Residue sequence	Identification expression	Substitution expression
QX^{1}P	(Q)([!P]P)	E\2
QX^{1}X(F,Y,W,M,L,I,V)	(Q)([!P]?[FYWMLIV])	E\2

X^1 refers to any amino acid with the exception of proline. For each sequence the appropriate identification and substitution expression are identified for use in Microsoft® Word

3.3 Prediction of Deamidation Sites Using Regular Expressions

1. Paste your query sequence into Microsoft® Word (*see* **Note 6**).

2. Open the Find dialog from the "Edit" menu.

3. Enter the identification expression that you wish to use (*see* Table 1) in the "Find" window (*see* **Note 7**).

4. Ensure that the "Use Wildcards" option is selected (Fig. 3). If this option is not visible you may need to select the "More" button to reveal extra options (*see* **Note 6**).

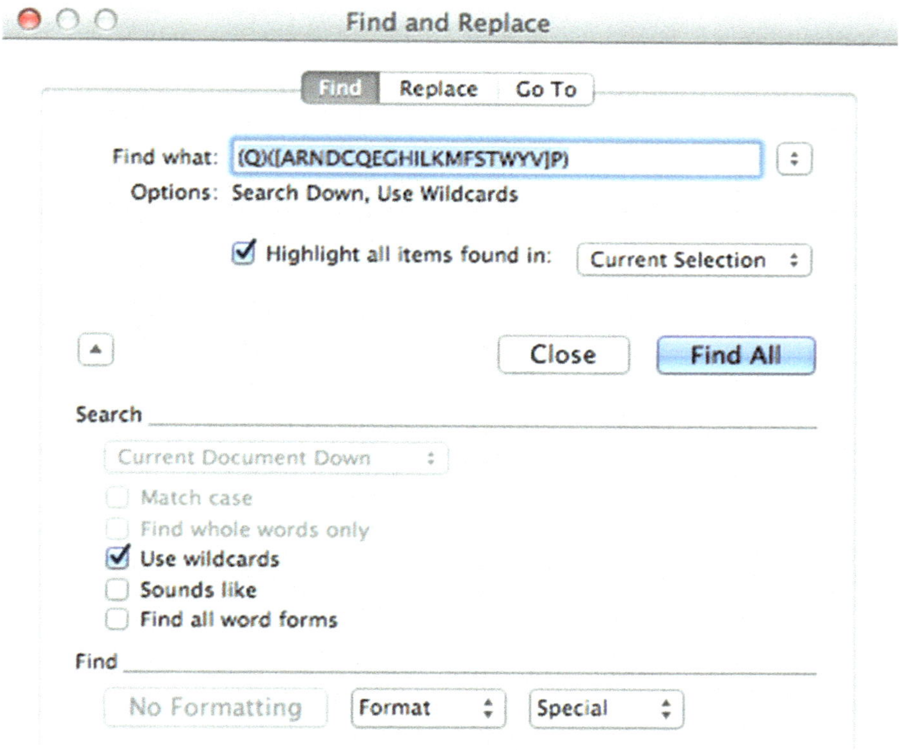

Fig. 3 The "Find" dialog in Microsoft® Word

5. Select the "Highlight all items found" checkbox and choose the find all option.

6. All of the potential deamidation sites will now be highlighted in your document.

7. If you wish to replace the residues in the text with their deamidated counterparts for further analysis it will be necessary to use the find and replace dialog. To do so select the "Replace" tab in the "Find and Replace" dialog box.

8. Leaving the selected "Find" expression in the "Find" field, enter the substitution expression as "E\2" (Fig. 4, *see* also **Note 8**).

9. Select "Replace All" to deamidate your sequence in silico. The locations of all deamidated residues will be highlighted in the text.

10. To save your deamidated sequence use the "Save as" dialog box and select "Plain text" as the format.

11. This sequence can be analyzed for predicted HLA-binding capacity using RANKPEP or another equivalent service.

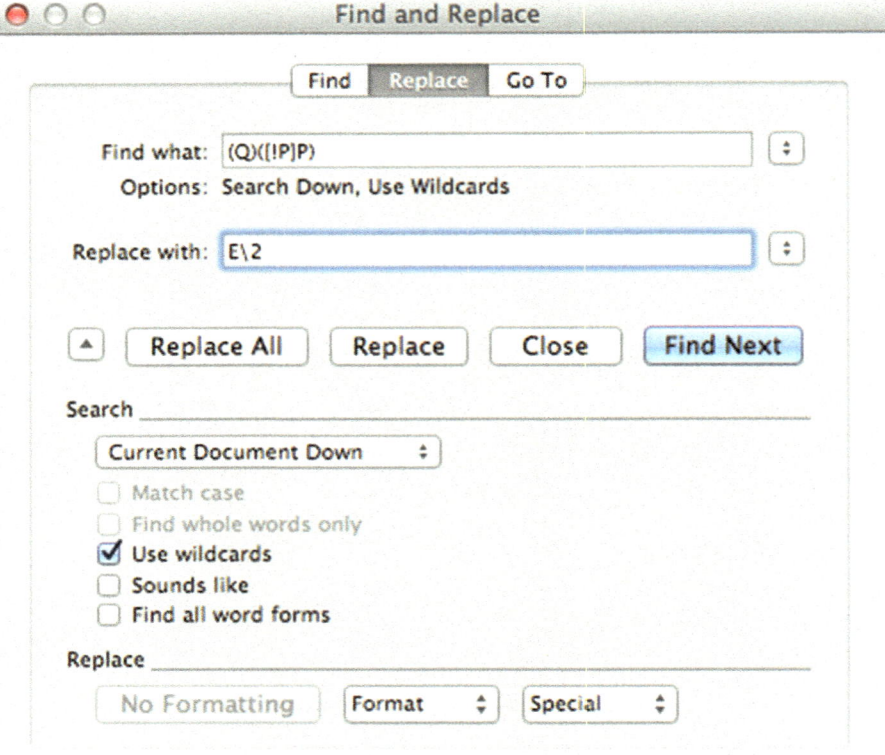

Fig. 4 The "Find and Replace" dialog in Microsoft® Word

4   Notes

1. The gliadin protein sequence with the accession number AGM38905.1 can be used to try the techniques from this chapter. The sequence can be obtained by searching the accession number via the UniProt webpage (http://www.uniprot.org).

2. The FASTA format is easily selected from a number of Web-based protein resources such as UniProt or the NCBI search engines. The FASTA format utilizes minimal annotation which is contained on a single line preceded by the character ">." The subsequent lines contain the protein sequence in single-letter amino acid code.

3. It is worth noting that in celiac disease the conversion of glutamine to glutamic acid is a characterized phenomenon due to the activity of the tissue transglutaminase enzyme. Thus, it may be wise to convert your sequences to their deamidated counterparts before conducting immunogenicity analysis (*see* **Subheading 3.3**).

4. Occasionally, you may experience trouble with "carriage returns" being copied into the query window. A carriage return

is a character that signifies that a new line should be commenced. This may mislead the program you are using into thinking it has reached the end of your protein sequence. Carriage returns can be deleted out manually by hitting the delete key at the end of a line that has a carriage return. For larger studies, it may be necessary to parse carriage returns out of your sequence automatically.

5. For digestive system enzymes, pepsin, trypsin, and chymotrypsin may represent a suitable starting point. However, depending on the biological system you wish to simulate you may need to change this list considerably. The PeptideCutter website contains brief descriptions of each of the enzymes that can be selected from the analysis page.

6. As many versions of Microsoft® Word exist, it would be beyond the scope of this chapter to detail the exact steps required for a given version of the software. The screenshots are taken from the Word for Mac 2011 software. Assistance for your particular software version can be found through the Microsoft Office help files or online help.

7. While a complete treatment of regular expressions is beyond the scope of this chapter we describe how each component of the regular expression operates. Brackets "()" divide the regular expression into two parts; the first part represents the Q residue of interest and the second represents the adjacent residues moving from the amino to the carboxy terminus of the protein. Each letter can be represented by a single character, e.g., Q for glutamine. ? represents any single character, [!P] denotes any single character except "P." The regular expression (Q)([!P]P) will highlight any location that has the letter Q followed by any letter except P followed by the letter P.

8. (Also refer to **Note** 7.) For the replacement entry "E\2," the letter "E" replaces the first item in brackets for your search. The "\2" tells the search algorithm to insert the second item in brackets after the letter "E." As the second item in brackets from your "Find" query will be the letters immediately after the letter "Q" this will result in just the letter "Q" being changed to "E."

References

1. Shan L, Molberg Ø, Parrot I et al (2002) Structural basis for gluten intolerance in celiac sprue. Science 297:2275–2279. doi:10.1126/science.1074129

2. Lin HH, Zhang GL, Tongchusak S et al (2008) Evaluation of MHC-II peptide binding prediction servers: applications for vaccine research. BMC Bioinformatics 9(Suppl 12):S22. doi:10.1186/1471-2105-9-S12-S22

3. Reche PA, Glutting J-P, Zhang H, Reinherz EL (2004) Enhancement to the RANKPEP resource for the prediction of peptide binding to MHC molecules using profiles. Immunogenetics 56:405–419. doi:10.1007/s00251-004-0709-7

4. Gasteiger E, Hoogland C, Gattiker A, et al (2005) The proteomics protocols handbook. doi:10.1385/1592598900

5. Vader LW, de Ru A, van der Wal Y et al (2002) Specificity of tissue transglutaminase explains cereal toxicity in celiac disease. J Exp Med 195:643–649

Chapter 17

Quality Control Procedures for High-Throughput Genetic Association Studies

Ciara Coleman, Emma M. Quinn, and Ross McManus

Abstract

Genome-wide association (GWA) studies provide an unbiased approach to discovering the role of genetic determinants of disease across the human genome. The case–control design, the most frequently used GWA study design employed to date, compares allele frequencies in affected patients to those of unaffected controls. Several large-scale GWA studies have identified numerous risk variants for celiac disease (CD). However, due to their low marker density, the early GWA arrays failed to adequately capture much of the genetic variance associated with CD. The Immunochip, a custom Illumina Infinium high-density array containing 196,524 common and rare polymorphisms, was developed to allow deep replication and fine mapping of the previously established GWA significant loci identified in 12 major autoimmune and inflammatory diseases, including CD. It has the advantage of allowing uniform sets of genetic markers to be compared across all diseases. This chapter describes the methods used to perform Immunochip genotyping and the bioinformatics steps necessary for quality control and analysis of the resulting data.

Key words Genome-wide association study, Immunochip, Celiac disease

1 Introduction

The National Institutes of Health defines a GWA study as a "study of common genetic variation across the entire human genome designed to identify genetic associations with observable traits" [1]. Thus far GWA studies have been carried out using population-based prospective, cross-sectioned, and case–control study designs [2]. The case–control design is the most frequently used GWA study employed to date. This type of study involves comparing allele frequencies between patients with the disease of interest to a group of unaffected individuals. GWAS have proven to be an important tool by allowing the correlation between disease status and genetic variation to be determined, thereby identifying candidate genes or specific genomic regions that contribute to disease. The GWA approach is revolutionary and powerful, as it provides an unbiased approach (as no prior biological

Anthony W. Ryan (ed.), *Celiac Disease: Methods and Protocols*, Methods in Molecular Biology, vol. 1326,
DOI 10.1007/978-1-4939-2839-2_17, © Springer Science+Business Media New York 2015

knowledge is required) to discovering the role of genetic determinants of disease across the entire human genome. While not inexpensive, it is sufficiently economical to allow thousands of unrelated individuals to be genotyped at high density in a manner that is required to detect the small effect sizes of typical "common" disease susceptibility alleles.

The first large GWAS was carried out by the Welcome Trust Case–Control Consortium (WTCCC) in 2007 [3], focusing on seven common diseases including coronary artery disease (CAD), rheumatoid arthritis (RA), and type 1 diabetes. Since then, many independent GWAS have been undertaken in a wide range of diseases. Among these, GWAS have been applied to the identification of non-HLA celiac disease (CD) risk loci in recent years. The first GWA study to be performed in CD was carried out by Van Heel et al. in 2007. They tested 310,605 SNPs for association in 778 patients and 1,422 controls. Outside the HLA region they identified CD risk variants in the 4q27 region containing the IL2 and IL21 genes [4]. Hunt et al. carried out a follow-up study in 2008 [5] where they genotyped 1,020 of the most strongly associated non-HLA markers in an additional 1,643 cases and 3,406 controls. This led to the identification of seven previously unknown risk variants, *IL12A*, *IL18RAP*, *RGS1*, *SH2B3*, *TAGAP*, *CCR3*, and *LPP*. A second GWA study published by Dubois et al. in 2010 [6] revealed an additional 13 new CD risk loci, *TNFRSF14*, *RUNX3*, *PLEK*, *CCR4*, *CD86*, *BACH2*, *PTPRK*, *ZMIZ1*, *ETS1*, *SOCS1*, *ICOSLG* and two regions with unidentified genes, bringing the total number of non-HLA CD susceptibility loci identified to 26.

However, early GWAS arrays were of low density, capturing limited variation across the genome and in particular in associated regions. As with most genome arrays, they were also heavily skewed towards common variation, particularly single-nucleotide variants. Thus the role of rare variants and others such as copy number variants could not be determined with these tools, and given the low density of markers, only a limited amount of genetic variation in regions of disease association was assayed, leading to low-resolution signals. In an effort to address these problems, the Immunochip genotyping array was developed to allow deep replication and fine mapping of previously established GWAS significant loci identified in 12 major autoimmune and inflammatory diseases including CD [7]. It has the added advantage of allowing uniform sets of genetic markers to be used across all diseases, facilitating cross comparison. The Immunochip is a custom Illumina Infinium high-density array that contains 196,524 common and rare polymorphisms (718 small insertion deletions and 195,806 SNPs) designed to perform dense genotyping of previously established GWAS significant loci from the major autoimmune and inflammatory diseases. These SNPs were selected by the consortium across 186 loci which reached genome-wide significance ($P < 5 \times 10^{-8}$) from

12 diseases (Crohn's disease, ulcerative colitis, type 1 diabetes, rheumatoid arthritis, psoriasis, systematic lupus erythematosus, primary biliary cirrhosis, ankylosing spondylitis, autoimmune thyroid disease, IgA deficiency, multiple sclerosis, and celiac disease) [7]. Wildcard SNPs from the 12 disease research consortia and SNPs from non-immunological diseases identified by the WTCCC2 were also included on the Immunochip [7]. All identified SNPs at the associated loci were included to maximize its utility for fine mapping, using sources such as the 1000 Genomes Project, repositories such as dbSNP, and other sequencing/variation data available to the consortium [8]. Immunochip v1 has ceased production and is due to be replaced in Q3 2015 by Immunochip v2 with approximately 2,75,000 SNPs, consisting of 1,80,000 v1 markers and 95,000 new markers suggested by the immunogenetics research community (Illumina, pers comm).

Immunochip analysis has been successful in identifying novel risk loci for a number of complex diseases, including primary biliary cirrhosis (PBC) [9], autoimmune thyroid disease (AITD) [10], bipolar disorder [11], and celiac disease [8]. In 2011, 12,041 individuals with CD and 12,228 controls were densely genotyped in the study by Trynka et al. They reported 57 independent CD association signals from 39 separate non-HLA loci, identifying 13 new CD risk loci reaching genome-wide significance ($P < 5 \times 10^{-8}$), bringing the total number of CD loci to 40 including the HLA locus. We have conducted an Immunochip analysis of an Irish CD case–control group and combining our results with those of Trynka et al. has led to the confirmation of two further loci (with likely candidate genes *ZNF335* and *NIFA*) as genome-wide significant bringing the total number of CD loci to 42 [12]. However, although much progress has been made, much of the genetic heritability of CD still remains unexplained. To date both the HLA and non-HLA loci combined are estimated to account for approximately 54 % of the genetic heritability of CD [13]. It is suggested that common variants of small effects and/or highly penetrant rare mutations have yet to be identified which may explain some or all of the remaining missing heritability.

In summary, GWAS have provided a hypothesis-free method of associating genetic variants with disease and the hunt for genes contributing to celiac disease has been particularly successful with 42 genome-wide significantly associated loci identified to date. The identification of such susceptibility loci is essential in aiding our understanding of the biological pathways involved in complex diseases and ultimately for uncovering new therapeutic targets, improved risk prediction, and personalized therapy.

This chapter describes the crucial quality assessment and quality control (QC) steps carried out during a typical case–control study using the Immunochip, an Illumina Infinium high-density array; however these QC steps can also be used for the analysis of data from other microarray platforms.

2 Materials

2.1 Isolating DNA from Peripheral Blood Samples

1. 4 ml VACUETTE® Premium tubes, lavender K3E K3EDTA (Cruinn Diagnostics).
2. VACUETTE® Safety Blood Collection Set (Cruinn Diagnostics).
3. Autopure LS robotic workstation (Qiagen).
4. Autopure RBC Lysis solution (Qiagen).
5. Autopure Cell Lysis solution (Qiagen).
6. Autopure precipitation solution (Qiagen).
7. Autopure DNA hydration solution (Qiagen).
8. Autopure glycogen solution (Qiagen).
9. Autopure 100 % Isopropanol (Qiagen).
10. Autopure 70 % Ethanol (Qiagen).
11. Autopure Qubes D (Qiagen).
12. Autopure Qubes E (Qiagen).
13. Heating block.
14. Orbital shaker.
15. Centrifuge.
16. 1.5 ml Eppendorfs.
17. TE buffer.

2.2 Preparation of 96-Well Plates

1. 96-Well plate.
2. Purified DNA.
3. Deionized H_2O.

2.3 Genotyping

1. Immunochip high-density array.
2. Illumina Infinium II protocol.
3. Illumina GenomeStudio GenTrain 2.0 software.

2.4 Data Quality Control

1. Raw GWA SNP data.
2. Computer workstation with Unix/Linux operating system.
3. PLINK software.
4. R statistical software.

   The freely available statistical software PLINK that is widely used for analyzing GWA data along with the statistical package R is used in this protocol.

3   Methods

3.1   DNA Isolation from Peripheral Blood Samples

The first step is to isolate DNA from whole blood samples.

1. Follow detailed instruction in the Autopure LS user manual (other DNA isolation techniques are available and may be used).

2. Samples can be stored at −20 °C until further use.

3.2   Preparation of 96-Well Plate

1. Add 50 ng of DNA to each well of a 96-well plate.

3.3   Genotyping

1. Individual DNA samples are genotyped for 1,96,524 genetic variants using the Illumina Infinium high-density array [13]. Bead intensity data are processed and normalized and genotypes are called using the supplied Illumina GenomeStudio GenTrain 2.0 software.

3.4   Data Quality Control (QC)

Due to the large quantity of marker loci tested in GWA studies, even small sources of systematic/random error or bias can give rise to erroneous results. Therefore, careful attention to data quality control (QC) is an essential component of all GWA studies. To maximize the numbers of markers in the study, it is recommended to begin the QC by removing individuals with particularly high error rates prior to conducting QC on a "per-marker" basis. The impact of removing one marker from a study is potentially greater than the removal of one individual as each marker removed is a possible overlooked disease association.

3.4.1   Cluster Plots

Intensity cluster plots generated by the Immunochip platform are inspected manually using Illumina's GenomeStudio Data Analysis Software to ensure that there are no clustering errors with any genotype calls before proceeding with SNPs for follow-up genotyping. Markers displaying poor clustering (Fig. 1) are removed. An example of successful clustering is shown in Fig. 2. Confirm that all of the remaining markers are of high quality.

3.4.2   Creation of BED Files

Genotype data are returned in various formats. Data can be exported from GenomeStudio in PLINK format, containing the standard PED and MAP file formats. A PED file is a delimited file where each line represents one individual and the first six columns are mandatory and contain "Family ID," "Individual ID," "Paternal ID," "Maternal ID," "Sex," and "Phenotype." Each subsequent column denotes genotypes. 0 signifies a missing genotype. Each SNP must have two alleles. A MAP file contains the order of the SNPs where each line represents a single marker and the columns are "Chromosome," "Marker name," "Genetic distance in

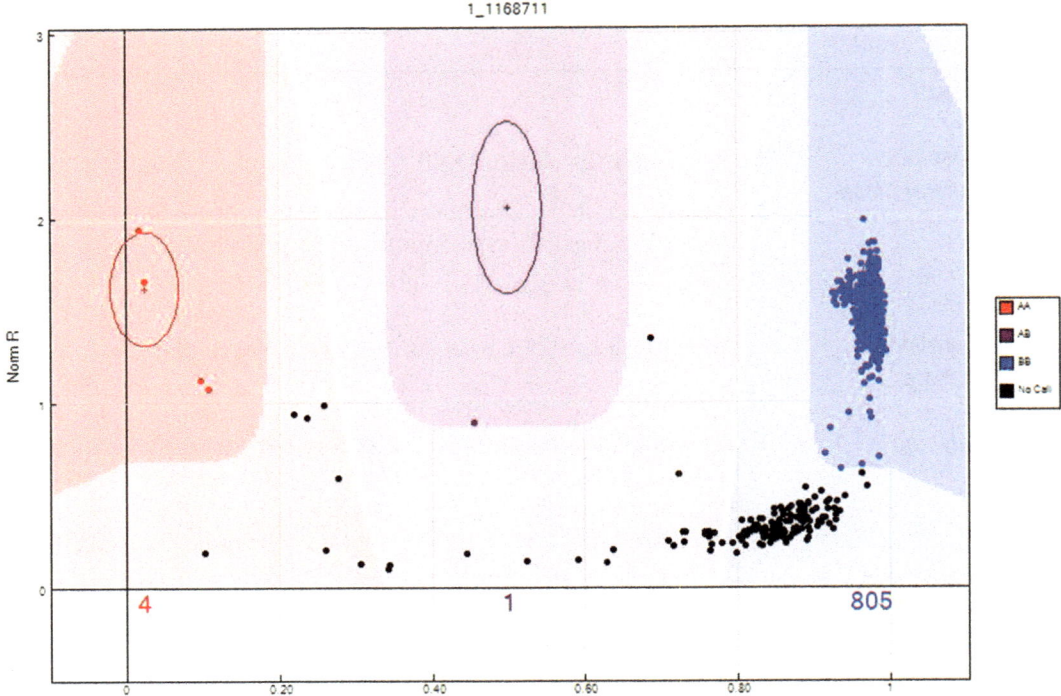

Fig. 1 Unsuccessfully clustered SNPs form diffuse clusters and do not correlate well with the standard cluster positions. These samples must be excluded

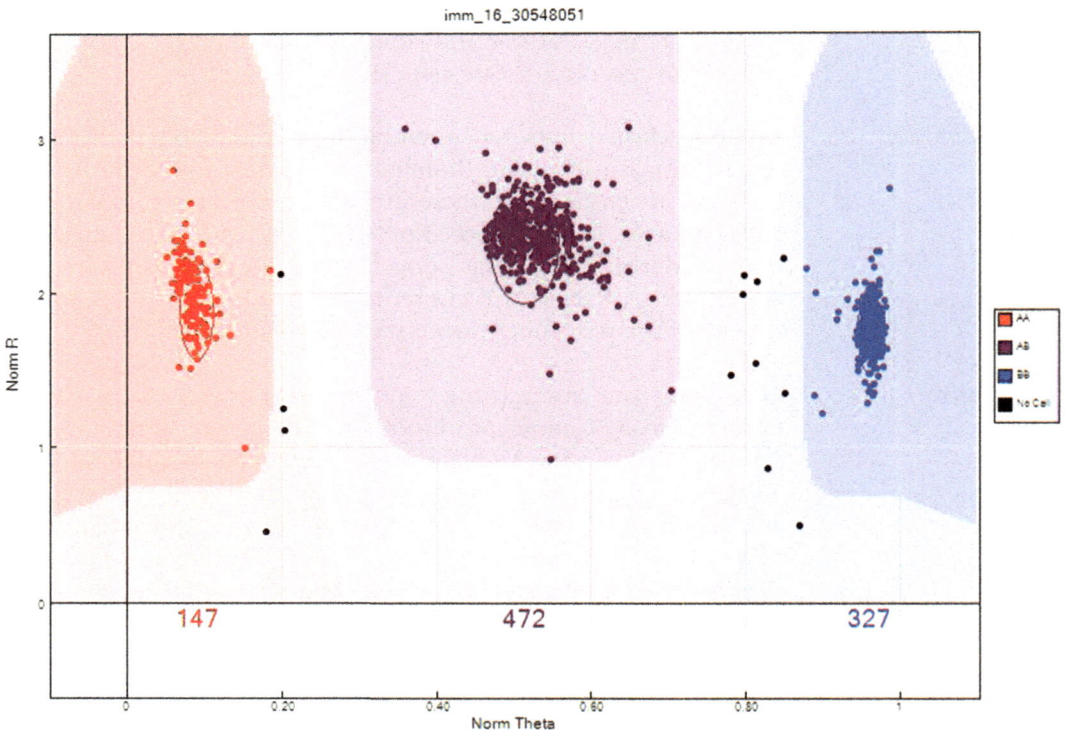

Fig. 2 Successfully clustered SNPs demonstrating three distinct genotypes aa, Aa, and AA

Morgans," and "Base-pair position." To assist the analysis of large-scale datasets, ped and map files may be converted to binary formats (bim, bed, and fam) (*see* **Note 1**).

1. Export the file from genome studio to create the raw files .ped and .map.

2. Binary file formats such as BED, BIM, and FAM files can be generated by typing

```
$ plink --file File_name --out File_
name --make-bed
   at the shell prompt.
```

3.5 Per-Individual Quality Control

Individuals are excluded based on five main criteria: their per-sample call rate, excessive heterozygosity, ancestry, duplication, and being closely related to another sample in the study, as described below.

3.5.1 Call Rate/ Missingness

Large variations can occur in DNA sample quality, which can have significant effects on genotype efficiency (call rate) and genotype accuracy; therefore individuals with systematically low call rates must first be excluded (*see* **Note 2**). The sample call rate is described as the fraction of called SNPs per sample over the total number of SNPs in the dataset.

1. At the shell prompt type

```
$ plink --bfile File_name --out File_
name --mind 0.1
```

(Mind > 0.1 implies that we accept individuals with less than 10 % missingness).

This will create the files .imiss and .lmiss. The N_MISS column in the .imiss file represents the number of missing SNPs whilst the F_MISS column shows the proportion of missing SNPs per individual.

3.5.2 Heterozygosity

The proportion of heterozygous loci in an individual (ranging from 0 to 1.0) is known as the heterozygosity. Any individuals showing high levels of heterozygosity should be removed from the dataset (*see* **Note 3**). Individuals with an excessive proportion of heterozygote genotypes may be an indication of DNA sample contamination. Alternatively individuals with a reduced proportion of heterozygote genotypes may be indicative of inbreeding.

At the shell prompt type

```
$ plink --bfile File_name --het --out
File_name
```

This command will create the file .het, where the third column contains the observed number of homozygous genotypes [O(Hom)] and the fifth column comprises of the number of non-missing genotypes [N(NM)] per individual. The heterozygosity rate per individual can then be calculated using the formula

$$Het = \left[N(NM) \right] \ O(Hom) / N(NM).$$

2. A graph can then be created where the observed heterozygosity rate per individual is plotted on the *x*-axis and the proportion of missing SNPs per individual is plotted on the *y*-axis. This can be created using standard software such as Excel or R.

3. These graphs can then be examined to identify thresholds at which to exclude individuals based on their SD from the mean heterozygosity rate.

4. Outlying samples can be removed by copying their family ID (FID) and individual ID (IID) into a text file (.txt) and by using the following command to remove them:

```
$ plink --bfile File_name --remove file.txt
```

5. Make a new BED file excluding outliers:

```
$ Plink --bfile File_name --out File_
name --make-bed
```

3.5.3 Sex Check

Genetic gender (based on heterozygosity rates of X chromosome SNPs) is compared to the gender reported in the PED file, and samples with mismatches should be removed.

1. At the shell prompt type

```
$ Plink --bfile File_name --check-sex
```

This generates a .sexcheck file.

2. Open this file and check for any inconsistencies; any samples that have discordant sex information can be removed by copying their family ID (FID) and individual ID (IID) into a text file (.txt) and by using the following command to remove them:

```
$ plink --bfile File_name --remove file.txt
--out File_name --make-bed
```

3.5.4 Identity by Descent

It is important in a standard population-based case–control association study that all the samples are unrelated. If samples are related or duplicate samples are included in the study, a bias is introduced whereby genotypes within families will be overrepresented, and therefore the sample may no longer be an accurate reflection of the allele frequencies in the entire population. To identify duplicate and related individuals, a metric (identity by state, IBS) is calculated for each pair of individuals based on the average proportion of alleles shared in common at genotyped SNPs (excluding sex chromosomes). Only independent SNPs are included in this analysis; regions of extended linkage disequilibrium (LD) are removed from the dataset and the remaining regions are pruned so that no pair of SNPs within 50 kb has an r^2 greater than 0.2. The --genome option in PLINK calculates pairwise kinship estimates between every individual in the study using a subset of markers. It also calculates the proportion of loci where two individuals share zero, one, or two alleles (relatedness). Duplicate individuals or monozygotic twins are defined as having an identity by descent (IBD) score of 1, IBD = 0.5 for first-degree relatives,

IBD = 0.25 for second-degree relatives, and IBD = 0.125 for third-degree relatives.

1. At the shell prompt type

```
    $ plink --bfile File_name --range \ --indep-
pairwise 50 5 0.2 --out File_name.prune.in
    $ plink --bfile File_name --extract File_
name.prune.in                              --
genome --out File_name
```

2. This command creates a .genome file which can be used to identify all pairs of individuals with pi-hat (measure of relatedness between pairs of individuals) >0.185. A score of >0.185 is halfway between the expected IBD for second- and third-degree relatives. These groups of individuals can then be manually examined to decide who should be excluded from downstream analysis. One individual from each pair should then be removed from the data.

3. A graph showing the proportion of loci sharing one allele (Z1 column of .genome file) can be plotted against the proportion of loci sharing zero alleles (Z0 column of .genome file) and plots color coded by relationship type using statistical software such as R.

3.5.5 Principal Component Analysis

Identifying and removing population stratification in the dataset is a critical part of individual QC, as it could cause spurious associations due to differences in ancestry rather than association to disease. Therefore, the next step is to identify and remove individuals of divergent ancestry and to define a homogeneous population for downstream analysis. Principal component analysis (PCA) evaluates the data by its underlying structure, i.e., the direction where there is the most variance. The first principal component explains as much variance in the data as possible with the subsequent principal components describing as much of the remaining variability as possible. When detecting ancestry, a principal component model is built using genome-wide genotype data from populations of known ancestry, where the principal components are compared against reference samples of known ethnicities. PCA can be performed on the entire SNP set, a filtered set of SNPs, or a pruned SNP set. A typical reference sample used is the HapMap genotype dataset from Europe (CEU), Asia (CHB and JPT), and Africa (YRI) due to the large differences between these populations. Several other datasets are available and can also be used.

1. Create a new bed file excluding those SNPs which do not feature in the genotype data of the four original HapMap populations; that is, extract SNPs in HapMap file from your dataset. Remove these SNPs by copying their family ID (FID) and individual ID (IID) into a text file (.txt) and by using the following command to remove them:

```
    $ plink --bfile File_name --remove file.txt
```

2. Make a new BED file excluding outliers:

```
$ Plink --bfile File_name --out File_name
--make-bed
```

3. Merge HapMap genotypes with your SNP data (download from http://hapmap.ncbi.nlm.nih.gov):

```
$ Plink -bfile file_name -bmerge hapmap.bed hap-
map.bim hapmap.fam --make-bed --out merged_genotypes
```

4. This is the final set of .BED, .BIM, and .FAM files which contain data on the study population and the four HapMap populations.

5. To perform MDS analysis in PLINK, repeat the IBD .genome analysis at this point to determine how related the sample is, following the addition of the HapMap SNPs on the pruned dataset:

```
$ Plink --bfile merged_genotypes --genome
--out File_name
```

Plink can then be used to calculate the MDS scores for the first two principal components for all individuals in the dataset:

```
$ Plink --bfile merged_genotypes --read-
genome File_name.genome --cluster --mds-plot 2
--out File_namemds
```

This command will produce a file called File_name.mds

6. Statistical software such as R can then be used to create a diagram of the first two principal components using all individuals in your dataset. Columns 2 and 3 will contain the first and second principal components, respectively. Each individual can also be color coded according to their respective ethnicity using the pedigree information located in the ped file.

7. Thresholds can then be set on the first and second principal components so that only individuals who match the ancestral population of interest are included. Individuals with a score outside of the set threshold can be removed.

8. Eigenstrat is another free, open-source software that can be used to perform this type of PCA. It is used to detect and correct for population stratification in large sample sizes in a computationally efficient manner.

3.6   Per-Marker QC

3.6.1   Missing Genotypes/Call Rate

Substandard markers must be removed from the data as they may result in false positives and reduce the ability to identify true associations correlated with disease risk. Typically, a call rate of 95 % and a threshold of ($P < 10e^{-6}$) for SNPs with differential missingness between cases and controls ($P < 10e^{-6}$) are removed from the dataset (*see* **Notes 4** and **5**).

1. At the shell prompt type

```
$ plink --bfile File_name --out File_name
--geno 0.05 --make-bed
$ plink --bfile File_name --out File_name
--test-missing
```

This command will create a .missing text file which contains the missing call frequencies of SNPs in both cases and controls. SNPs can be excluded based on a given threshold determined by their p-value (e.g., $P < 10e^{-6}$).

2. These samples can be removed by copying their family ID (FID) and individual ID (IID) into a text file (.txt) and by using the following command to remove them:

```
$ plink --bfile File_name -remove file.txt
-make-bed -out file_name
```

3.6.2 Hardy-Weinberg Equilibrium

Hardy-Weinberg equilibrium (HWE) is one of the most important principles in population genetics. Under Hardy-Weinberg assumptions, both allele and genotype frequencies remain constant from generation to generation. Therefore, population allele frequencies at given loci can be used to calculate the equilibrium-expected genotypic proportions. Departure from this equilibrium may represent inbreeding, population stratification, or genotyping errors. Control samples only are used, as it is assumed that there is a biological explanation for the deviation at particular loci in patient samples. Markers that show deviations from HWE at loci associated with disease in control samples ($P < 10e^{-6}$) should be removed from the dataset:

```
$ plink --bfile File_name --out File_name
--hwe 0.00001 --make-bed
```

3.6.3 Minor Allele Frequency

The removal of all SNPs with a very low minor allele frequency (MAF) is the final step in the QC process. It is important to remove these SNPs as the statistical power is very low for rare SNPs. Calling rare SNPs using current genotype calling algorithms can be difficult due to the small size of the heterozygote and rare homozygote clusters and can present as false positives in case–control association tests. The removal of rare SNPs does not have a huge impact on the study, as the power to detect an association at rare variants is so low. The --maf option will remove all SNPs with a MAF less than a specified threshold (*see* **Note 6**). A threshold of 0.05 is used in the example shown here:

```
$ plink --bfile File_name --out File_name --maf
0.05 --make-bed
```

3.7 Association Testing

After assuring the quality of the data, association testing is performed to test for a correlation between disease status and genetic variation.

1. To perform a standard case–control association analysis type

```
$ plink --bfile File_name --assoc --out File_name
```

2. To perform logistic regression association analysis type (*see* **Note 7**)

```
$ plink --bfile File_name --logistic --out File_name
```

3. To perform further tests of association between a disease and a variant other than the basic allelic test type

```
$ plink --bfile File_name --model --out
File_name
```

4   Notes

1. The creation of BED files stores data in a more memory-efficient manner to assist with the analysis of large datasets.

2. Genotyping thresholds can vary between studies—common thresholds are >0.05 and >0.1.

3. A typical heterozygosity threshold is to remove individuals with a heterozygosity rate of ±3 standard deviations (SD) from the mean.

4. Carefully examining the distribution of missing genotype rates/distribution of mean heterozygosity across the entire sample set is the best way to establish the appropriate thresholds to use.

5. Typically markers with a call rate less than 95 % are removed from most studies. However, some studies have chosen higher call rate thresholds (99 %) for markers of low frequency (MAF <5 %).

6. Normally a MAF threshold of 1–2 % is applied but studies with a smaller sample size may need to set the threshold higher, i.e., 5 %.

7. Multiple covariates can be included in the regression model depending on the study being performed.

Acknowledgments

We would like to thank Dr. Anthony W. Ryan, Ms Valerie Trimble, and all who contributed samples for their help. This work was supported by Science Foundation Ireland (SFI) grant 09/IN.1/B2640 to R.M.M.

References

1. National Institutes of Health (2007) Policy for sharing of data obtained in NIH supported or conducted genome-wide association studies (GWAS). Federal Regist 72(166):49290–49297. http://www.grants.nih.gov/grants/guide/notice-files/NOT-OD-07-088.html. Accessed on January 2015

2. Turner S, Armstrong LL, Bradford Y et al (2011) Quality control procedures for genome wide association studies. Curr Protoc Hum Genet. Chapter 1:Unit 1.19

3. Burton PR, Clayton DG, Cardon LR et al (2007) Genome-wide association study of 14,000 cases of seven common diseases and 3,000 shared controls. Nature 447:661–678

4. Van Heel DA, Franke L, Hunt KA et al (2007) A genome-wide association study for celiac disease identifies risk variants in the region harboring IL2 and IL21. Nat Genet 39:827–829

5. Hunt KA, Zhernakova A, Turner G et al (2008) Newly identified genetic risk variants for celiac disease related to the immune response. Nat Genet 40:395–402

6. Dubois PCA, Trynka G, Franke L et al (2010) Multiple common variants for celiac disease influencing immune gene expression. Nat Genet 42:295–302

7. Cortes A, Brown MA (2011) Promise and pitfalls of the Immunochip. Arthritis Res Ther 13:101

8. Trynka G, Hunt KA, Bockett NA et al (2011) Dense genotyping identifies and localizes multiple common and rare variant association signals in celiac disease. Nat Genet 43:1193–1201

9. Juran BD, Hirschfield GM, Invernizzi P et al (2012) Immunochip analyses identify a novel risk locus for primary biliary cirrhosis at 13q14, multiple independent associations at four established risk loci and epistasis between 1p31 and 7q32 risk variants. Hum Mol Genet 21:5209–5221

10. Cooper JD, Simmonds MJ, Walker NM et al (2012) Seven newly identified loci for autoimmune thyroid disease. Hum Mol Genet 21:5202–5208

11. Green EK, Hamshere M, Forty L et al (2013) Replication of bipolar disorder susceptibility alleles and identification of two novel genome-wide significant associations in a new bipolar disorder case–control sample. Mol Psychiatry 18:1302–1307

12. Coleman C, Quinn EM, Ryan AW, et al (2015) Common polygenic variation in coeliac disease and confirmation of ZNF335 and NIFA as disease susceptibility loci. Eur J Hum Genet. doi:10.1038/ejhg.2015.87

13. Romanos J, Rosén A, Kumar V et al (2014) Improving coeliac disease risk prediction by testing non-HLA variants additional to HLA variants. Gut 63:415–422

Chapter 18

Quality Control and Analysis of NGS RNA Sequencing Data

Emma M. Quinn and Ross McManus

Abstract

Transcriptome sequencing, where RNA is isolated, converted to library of cDNA fragments, and sequenced using next-generation sequencing technology, has become the method of choice for the genome-wide characterization of mRNA levels. It offers a more accurate quantification of transcript levels than array-based methods, but also has the added benefit of allowing the discovery of novel gene/transcripts, alternative splice junctions, and novel RNAs. In addition, RNA sequencing may be used to investigate differential gene expression, allelic imbalance, eQTL mapping, RNA editing, RNA-protein interactions, and alternative splicing. A number of statistical methods and tools are available for differential expression analysis using RNA sequencing data and these are continually being developed and improved to handle more complex experimental designs. This chapter describes an example workflow for the quality control and analysis of raw RNA sequencing reads for the purposes of differential gene expression analysis, followed by pathway/enrichment analysis of significantly different genes. The methods and tools described are just one example of how this analysis can be conducted, but they can be applied to most standard RNA sequencing studies of differential gene expression. The methods covered are based on Illumina HiSeq single-end 50 bp reads. However, all programs used are capable of working with paired-end data, subsequent to minor adaptations.

Key words Celiac disease, RNA sequencing, Transcriptome

1 Introduction

In recent years, transcriptome sequencing has become the method of choice for the genome-wide characterization of mRNA levels as it not only offers a more accurate quantification of transcript levels in a given cell or tissue but also has the added benefit of allowing the discovery of novel gene/transcripts, alternative splice junctions, and novel RNAs [1–3]. In addition the applications of RNA sequencing analysis stretch further than characterization of the transcriptome to differential gene expression, allelic imbalance [4, 5], eQTL mapping [6], RNA editing [7], RNA-protein interactions [8], and alternative splicing [9, 10]. The principle and RNA sequencing process has been described elsewhere but in brief; RNA is isolated from the source of choice, converted to library of cDNA

Anthony W. Ryan (ed.), *Celiac Disease: Methods and Protocols*, Methods in Molecular Biology, vol. 1326,
DOI 10.1007/978-1-4939-2839-2_18, © Springer Science+Business Media New York 2015

fragments, and sequenced using next-generation sequencing technology to obtain short sequences or "reads" (either single end or paired end). These reads can then be aligned or mapped to a reference genome if available or used for the de novo assembly of genes or transcripts.

When designing an RNA Seq experiment for differential expression analysis, there are a number of factors that are important to consider: (1) The experimental design itself—is it straightforward (i.e., comparing gene expression between two conditions, e.g., disease vs. control) or more complex (including different time points, drug dosages, genotype)? (2) Sequencing depth—the number of reads required depends on the application; for example a great many more reads are necessary for transcriptome characterization in order to ensure that the transcriptome is captured in its entirety than is required for differential expression analysis. Studies have shown that as little as 10 million reads can be sufficient to quantify gene expression across samples: however, it depends on the effect sizes you wish to observe [11, 12]. (3) Number of replicates— due to the high costs involved with RNA Seq it is sometimes the case that studies are performed using little or no biological replication. However, as the technology becomes more widely used for differential expression analysis the consensus is that in order to make any biologically meaningful interpretation of the data, at least three replicates of each group being analyzed should be performed in order to properly assess the natural biological variation within each. In fact it has been shown that increasing the number of replicates far improves the power of the study over increasing the number of reads or sequencing depth [13]. In order to assess the number of samples and depth best suited to your data, it may be beneficial to assess your data through a power calculator tool when designing your experiment [14, 15].

There are a number of statistical methods and tools available for differential expression analysis using RNA sequencing data and these are continually being developed and improved to handle more complex experimental designs as RNA sequencing becomes more affordable and widely used. Once aligned, RNA sequencing reads can be quantified in the form of fragments per kilobase of transcript per million mapped reads (FPKM), which takes into account gene length and the total number of mapped reads or by simply counting the number of reads at a given locus (e.g., gene or exon boundary). FPKM is generally considered most useful when comparing gene expression within a given sample and for the analysis of gene expression between samples, the majority of the most commonly used tools require raw read counts as used in this chapter.

This chapter describes an example workflow for the quality control and analysis of raw RNA sequencing reads for the purposes of differential gene expression analysis, followed by pathway/enrichment analysis of significantly different genes. The methods and tools described and used throughout are just one example of how this analysis can be conducted but should be able to be applied to most standard RNA sequencing studies of differential gene expression (*see* **Note 1**). Further reading of the vignettes and manuals for the various programs used in this chapter is advised for more complicated experiments (*see* **Note 2**). The methods covered in this chapter are based on human RNA sequenced on an Illumina HiSeq using single-end 50 bp reads; however all programs used are capable of working with paired-end data subsequent to minor adaptations.

2 Materials

2.1 Computing Resources

64-bit CPU computer running on Linux with a minimum of 4 Gb RAM (preferably 16 Gb) advised. Depending on the number of samples involved hundreds of Gb may be required for storage. Basic knowledge of the command line and R is assumed.

2.1.1 Software

Download and install the programs:

FastQC; http://www.bioinformatics.babraham.ac.uk/projects/fastqc/

TopHat2 http://tophat.cbcb.umd.edu/

SAMtools http://samtools.sourceforge.net/, R http://www.r-project.org/

Trimmomatic http://www.usadellab.org/cms/?page=trim-momatic

3 Methods

3.1 Quality Control Assessment of Raw Fastq RNA Seq Reads

Post-sequencing, the data from an RNA Seq experiment conducted on an Illumina sequencing machine will more than likely be presented in FASTQ format.

In order to take a quick look at your files open a Linux terminal and navigate to the folder containing .fastq files:

```
$ head <myfastqfile>.fastq #will show the first
10 lines of your file
```

Each read in your fastq file contains four lines. The first line contains the sequence identifier, the second contains the sequence itself in fastq format, and the third contains a "+" symbol which signals the end of the sequence and start of the quality string. The fourth line contains the Ascii-encoded base quality scores for each base in the read.

A simple way to identify how many reads are in your fastq file is to count the number of lines in the file and divide by four (*see* **Note 3**):

```
$ wc -l <myfastqfile>.fastq | awk '{print $1/4}' #
will display the number of reads in "myfastqfile.fastq"
```

3.1.1 Quality Control of Fastq Files

The number of sequencing reads will be too great to visualize and assess manually, so several tools exist in order to conduct this efficiently.

There are several open-source software tools, e.g., RSeQC [16], and R packages, e.g., Short Read [17], available to assess the quality of RNA sequencing reads and identify any abnormalities. Here we will use the tool FastQC to examine the raw fastq data.

FastQC is Java software with a graphical interface that works with data in FASTQ, BAM, or SAM format. It creates a report in HTML format with summary graphs and tables that should help assess sequence data. It can either run as a stand-alone interactive application for the immediate analysis of small numbers of FastQ files or be integrated into a larger analysis pipeline for the systematic processing of large numbers of files.

Once installed, FastQC can be run by entering the following on the command line:

```
fastqc  --help # to obtain a list of available
options
fastqc -q* <myfastqfile>.fastq # to run FastQC
```

-q will supress all progress messages and only report errors.

FastQC evaluates the data by running a series of analysis modules which include Basic Statistics, per-base sequence quality, per-base sequence content, per-base GC content, per-sequence GC content, per-base N content, sequence length distribution, duplicate sequences, overrepresented sequences, overrepresented Kmers, duplication levels, and kmer profiles. The left-hand side of the html report will display for the given data whether the data seems normal (green tick), slightly abnormal (orange triangle), or unusual (red cross). These should be interpreted in the context of your own data and more details on each evaluation module can be obtained in the help section (*see* **Note 4**).

3.1.2 Base Trimming and Removing Adapter Sequences

At this stage, depending on your data and the output results from FastQC, it may be necessary to remove adapter sequences present in the data or to remove bases with a low-quality score. Again there are several tools available to conduct this including Cutadapt (https://code.google.com/p/cutadapt/), FASTX-toolkit (http://hannonlab.cshl.edu/fastx_toolkit/index.html), or the R package ShortRead [17]. Here we show how to perform these tasks using the tool Trimmomatic; the code below should be adjusted to the relevant step options for your data:

```
$ Java -jar <path-to-trimmomatic.jar>   org.
usadellab.trimmomatic.Trimmomatic.SE  [-phred64
| -phred33]* <myfastqfile.fastq> <output.fastq>
<options> ....
```

*refers to the quality scores used by Illumina and will be specific to your data (*see* **Note 5**).

Trimmomatic Options

ILLUMINACLIP:<fastaWithAdaptersEtc>:<seed mismatches>:<palindrome clip threshold>:<simple clip threshold>

- fastaWithAdaptersEtc: specifies the path to a fasta file containing all adapter sequences you wish to remove.

- seedMismatches: specifies the maximum mismatch count which will still allow a full match to be performed.

- palindromeClipThreshold: specifies how accurate the match between the two "adapter ligated" reads must be for PE palindrome read alignment.

- simpleClipThreshold: specifies how accurate the match between any adapter etc. sequence must be against a read.

SLIDINGWINDOW:<windowSize>:<requiredQuality>

- windowSize: specifies the number of bases to average across.

- requiredQuality: specifies the average quality required.

LEADING:<quality> #cuts bases off at start of a read

- quality: specifies the minimum quality required to keep a base.

TRAILING:<quality> #cuts bases off at end of a read

- quality: specifies the minimum quality required to keep a base.

CROP:<length> #cuts the reads to a specific length

- length: the number of bases to keep, from the start of the read.

HEADCROP:`<length>` #cut the specified number of bases from the start of a read

- length: the number of bases to remove from the start of the read.

MINLENGTH:`<length>`

- length: specifies the minimum length of reads to be kept.

For example for the fastq file "myfastqfile.fastq" encoded using phred 33 quality scores the following command would remove a list of adapters contained in the file adaptors.fa, remove leading and trailing bases with quality less than 3, scan the read with a 4-base wide sliding window, cutting when the average quality per base drops below 15, and drop any read that is below 36 bases long after trimming.

```
$ java -jar <path-to-trimmomatic.jar> SE --phred33
<myfastqfile.fastq>  <output.fastq>  ILLUMINACLIP:
Adapters.fa:2:40:15 LEADING:3 TRAILING:3 SLIDINGWIND
OW:4:15 MINLENTH:36
```

3.2 Sequence Alignment

Again there are a number of tools available to conduct sequence alignment for RNA Seq. Choice of tool will reflect the type of data involved, how well characterized the reference genome is, if there is one, whether you wish to align to the genome or transcriptome, etc. Alignment to the genome takes longer; however by just aligning to the transcriptome you can lose information on the presence of novel or uncharacterized mRNA splice variants.

For this analysis we use the splice aligner TopHat2 [18] which predicts transcripts from genome-mapped RNA Seq reads along with known and predicted splice junctions between exons. The reference and annotation files for a large number of commonly analyzed organisms that are compatible for TopHat2 can be downloaded from iGenomes http://support.illumina.com/sequencing/sequencing_software/igenome.ilmn for Ensembl, NCBI, or UCSC builds. It is also possible to build your own index files using Bowtie if these files are not available for your given organism. Here we give an example of commands used to align RNA Seq reads in fastq format to the UCSC human reference genome hg19.

Entering tophat2 into the terminal will give a list of the various parameters and settings that can be edited when running tophat along with the default values for each.

```
$ tophat2 [options ] <path to bowtie index>
<myfastqfile.fastq> #to run tophat2
```

3.2.1 Some of the Main Options to Consider

-**g**/ –**max-multihits**—A significant number of RNA Seq reads may map to more than one location leading to an over/underestimation of gene coverage. Discarding these reads can result in loss of information and potential underestimation of gene expression levels. However, it is common to set this to one so that only uniquely mapped reads are left for subsequent analysis.

-**G**/--**GTF**—Allows you to provide tophat with a set of gene models/annotations to aid mapping reads across exon junctions. Reads that do not fully map to the transcriptome will then be mapped on the genome.

-**T**/--**transcriptome-only**—This will allow you to only align the reads to the transcriptome and report only those mappings as genomic mappings.

-**N**/--**read-mismatches**—The number of mismatches allowed per read.

-**m**/--**splice-mismatches**—Number of mismatches allowed, maximum number of mismatches that may appear in the "anchor" region of a spliced alignment.

--**segment-length**—Specifies the length that each read will be cut into.

--**segment-mismatches**—Specifies the number of mismatches allowed per segment.

TopHat2 produces a number of files containing information on the junctions, insertions, and deletions identified during the alignment (junctions.bed, insertions.bed, deletions.bed) as well as reads for which a suitable alignment could not be found (unmapped. bam). The reads that have been aligned are contained in the file accepted_hits.bam and in the next section we will go through the steps necessary to assess the quality of the alignment.

3.2.2   QC of Aligned Sequences

The Picard suite of tools is a java-based program for the manipulation of files in .sam or .bam format. It can be used to perform tasks such as the marking of duplicate reads (which can subsequently be removed using SAMtools; see below), filtering, comparing, sorting, etc. Here we use Picard to collect information on the alignment of RNA to various functional classes of loci in the genome: coding, intronic, UTR, intergenic, and ribosomal.

Example command use to run the CollectRnaSeqMetrics program:

```
java -jar /CollectRnaSeqMetrics.jar REF_FLAT=
<path to refFlat.txt> I=accepted_hits.bam O=
myfilename_rnaseqmetrics.txt R=<path_to_hg19.fa>
STRAND=NONE
```

This will generate a file "filename_rnaseqmetrics.txt" that gives details on the number and percentage of aligned, coding, UTR, ribosomal, intronic, and intergenic bases. This can be useful when dealing with a number of samples to check for consistency and alignment rate.

3.2.3 Useful SAMtools Functions

Sort (*see* **Note 6**)

```
$ samtools sort accepted_hits.bam accepted_
hits_sorted. bam
```

Index (*see* **Note 6**)

```
$ samtools index accepted_hits.bam
```

Convert BAM to SAM

It may sometimes be necessary to convert files in BAM format to SAM for use with downstream tools (such as HTSeq-count (section 7)).

```
$ samtools view accepted_hits.bam  accepted_
hits.sam
```

Remove Duplicates (*see* **Note 7**)

```
$ samtools rmdup accepted_hits.bam accepted_
hits_dups_rm.bam
```

3.3 Summarizing Read Counts Mapped to Genes

In order to conduct differential gene expression analysis it is necessary to obtain a quantitative measure of the number of reads aligned within gene boundaries. Again there are several tools, e.g., Cufflinks [19], and R packages, e.g., easyRNASeq [20], that can be used for this. Here we use the tool HTSeq-count. Given a file with aligned sequencing reads and a list of genomic features (genes), HTSeq-count will produce a table of read counts for each gene annotated in the GTF file along with information on reads that were not counted for various reasons, e.g., reads that could not be assigned to any feature, ambiguous reads, or reads with more than one reported alignment. An example of the command use to run HTSeq-count is given below:

```
$ htseq-count -s* no sample1_accepted_hits.
sam <path to gtf> > sample1_accepted_hits_gene-
counts.txt
```

*-s signifies that the data is not from a stranded specific assay (default is yes).

Remove the last five lines of the HTSeq-count output file which lists the information on reads that were not counted.

```
$ head -n -5 sample1_accepted_hits_genecounts.txt
> sample1_genecounts_forDESeq.txt
```

In order to use the output from HTSeq in DESeq2, the count files should be merged for all samples to create a matrix of read counts (*see* **Note 9**). This can be carried out by pasting the output from HTSeq for each sample into a spreadsheet program such as Microsoft excel or using merge functions in Stata or R and saving the merged table as a .txt or .csv file. The resulting table should have the following format: a column for each sample and a row for each gene (*see* Table 1 below).

Table 1
Example count file layout for DESeq2

Gene ID	untreated_1	untreated_2	treated_1	treated_2
Gene 1				
Gene 2				

3.4 Differential Gene Expression Analysis

Here we use the program DESeq2 to normalize counts and test for differential gene expression. The following will list the steps necessary for the analysis of DE using DESeq2 version 1.2.8. Note that whilst it is possible to use DESeq2 to analyze data with no biological replicates, they are necessary in order to produce any biologically meaningful results.

DESeq2 runs within R and the following is an example of the code used to perform DE analysis according to the vignette for version 1.6.2. As the analysis produces a number of files and images it might be beneficial to create a new directory at this stage of the analysis.

```
$ mkdir DESeq #creates a new directory "DESeq"
$ mv merged_counts.txt DESeq # moves file containing merged counts for all samples to the DESeq folder
$ R # Launch R -from a unix terminal
> setwd ("<path-to-DESeq-directory>") #Set the working directory to the folder containing the countfile created in section 6.
> source("http://bioconductor.org/biocLite.R")
>biocLite("DESeq2") #To install the package DESeq2 enter (see Note 8)
> library("DESeq2")
```

3.4.1 Importing the Data (See Note 9)

Assuming that you have prepared a matrix of read counts from the files created using HTSeq section 6 the following will apply; for details of other methods of data input please see the latest DESeq2 vignette.

```
>countData=as.matrix(read.table("merged_counts.txt", header=TRUE, row.names=1))
>design=data.frame(row.names=colnames(countData), condition=c(rep(" untreated",2),rep("treated",2)))
> dds <- DESeqDataSetFromMatrix(countData = countData,colData=design,design=~condition)
> colData(dds)$condition <- factor(colData(dds)$condition,
  levels=c("untreated","treated"))
```

3.4.2 Data Quality Assessment by Sample Clustering and Visualization

For visualization and clustering purposes it is necessary to work with transformed versions of the count data. The following commands create a regularized log transformation and variance-stabilizing transformation of the data which can be used to create the heatmap of sample-to-sample distances and PCA plots below.

```
>rld <- rlog(dds, blind=TRUE)
>vsd <- varianceStabilizingTransformation
(dds, blind=TRUE)
>install.packactes("gplots") #(see Note 8)
>install.packages("RColorBrewer") #(see Note 8)
>library ("RColorBrewer")
>library ("gplots")
>distsRL <- dist(t(assay(rld))) #Create heat-
map showing the Euclidean distances between the
samples as calculated from the regularized log
transformation
>mat <- as.matrix(distsRL)
>rownames(mat) <- colnames(mat) <- with(coID
ata(dds),paste(condition,sep=" : "))
> hmcol <- colorRampPalette(brewer.pal(9,
"GnBu"))(100)
>pdf("RNA_seq_results.pdf")
>heatmap.2 (mat, trace="none", col = rev(hmcol),
margin=c(10, 10))
>print (plotPCA(rld,intgroup=c("condition")))
# PCA plot-useful for assessing batch effects
and outliers.
```

3.4.3 Differential Expression Analysis

```
>dds <- DESeq(dds) #this step performs all stages
of the DE analysis including normalisation and
dispersion estimation but it is also possible to
perform each test individually. see Note 10 and
the vignette for details.
>res <- results (dds)
>resOrdered <- res[order(res$padj),]#order
results by adjusted pvalue
>head (resOrdered) #view results
>nrow(res[res$padj<0.1 & !is.na(res$padj),])
# Number of DE genes at a FDR of 10%
>plotDispEsts(dds) # inspect estimated
dispersions
>plotMA(dds) # visualise differential expres-
sion (fold changes) V number of read counts
```

3.4.4 Create Results Tables

```
res<-na.omit(res)
write.csv (as.data.frame(res),file=" Results_
allGenes.csv") #create results table for all genes
>resSig <- subset(resOrdered, padj < 0.1)
```

```
>write.table(as.data.frame(resSig),file="Results_
SigGenes.csv",quote=FALSE, sep="\t", row.names=T) #
create results table for all genes significant using
an FDR 10%
```

3.4.5 Prepare Files for GOSeq Analysis (Section 8)

```
>allGenes=row.names(res)
>write.table (as.data.frame(allGenes),file=
"allGenes_goseq.txt",quote=FALSE, sep="\t", row.
names=FALSE) #creates list of all gene IDs tested
as part of the analysis
>sigGenes=row.names(resSig)
>write.table(as.data.frame(sigGenes),
file="resSig_for_goseq.txt",quote=FALSE,
sep="\t", row.names= FALSE) # creates list of
significantly DE genes
```

3.5 Pathway Analysis Using GOSeq

There are a number of different publically available tools and R packages for the pathway analysis of gene expression data, e.g., DAVID [21] and Piano [22]. Here we use the R package GOSeq to perform Gene Ontology analysis on our data; GOSeq takes into account and adjusts for the selection biases that occur in RNA Seq data, in that more highly expressed and/or longer transcripts offer more statistical power to detect differential expression and are therefore more likely to be detected as DE. Here we use the program to test for overrepresentation of GO terms only; however it is possible to use GOSeq to test for enrichment of additional pathways, e.g., KEGG or a custom user-defined pathway (*see* **Note 11**).

GOSeq requires a list of all genes that were tested as part of the RNA Seq experiment and those that are differentially expressed. The list of differentially expressed genes used in the code below is all genes that were DE—the file "resSig_for_goseq.txt" from the previous analysis (*see* **Note 12**).

```
>source ("http://bioconductor.org/biocLite.R")
>biocLite ("goseq")      #(see Note 8)
>library (goseq)
>all.genes <- scan ("allGenes_goseq.txt",what=
character(),skip=1)
>de.genes <- scan ("resSig_for_goseq.txt",
what=character (), skip=1)
>gene.vector = as.integer(all.genes %in% de.
genes)
>names (gene.vector)=all.genes
>pdf("GOseq.pdf")
>pwf=nullp(gene.vector, "hg19", "geneSymbol")
# Fit the Probability Weighting function for each
gene
>head (pwf) # view result
```

```
>GO.wall=goseq(pwf, "hg19", "geneSymbol")  #
calculate the over and under expressed GOcategories
  >head (GO.wall) # view results
  >write.table(GO.wall, file="GOtable.txt", quote=
FALSE, sep="\t", row.names=FALSE) #create results
table
  >enriched.GO=GO.wall$category[p.adjust(GO.
wall$over_represented_pvalue,method="BH")<.1]
# FDR 10% #introduce a FDR cut off of 10%
  >GO.enrichedFDR=p.adjust(GO.wall$over_
represented_pvalue,method="BH")
  >GO.enrichedFDRsig=GO.enrichedFDR<.1
  >go_results=data.frame(GOid=enriched.GO,FDR
=GO.enrichedFDR[GO.enrichedFDRsig])
  >write.table(go_results, file="GOenriched.txt",
quote=FALSE, sep="\t", row.names=FALSE, col.names=
TRUE) # create results table of GO terms signifi-
cantly enriched within specified FDR
```

3.5.1 Adding Information on the GO Terms to the Output

Further detail on each of the GO terms can be obtained through the R package GO.db.

```
> source ("http://bioconductor.org/biocLite.R")
> biocLite ("GO.db")     # (see Note 8)
>library(GO.db)
>sink (file="GOinfo.txt", type="output")
>for (go in enriched.GO[]){
+ print (GOTERM[[go]])
+ cat ("------------------------------------\n")
+ }
sink()
```

3.5.2 Performing Analysis on Individual GO Categories: Cellular Components, Biological Processes, and Molecular Functions

1. *Molecular Functions*

```
>GO.MF=goseq(pwf, "hg19", "geneSymbol", test.
cats=c("GO:MF"))
>enriched.MF=GO.MF$category[p.adjust(GO.MF$
over_represented_pvalue,method="BH")<.1]
>write.table  (enriched.MF,  file="GOenriched
MF.txt",  quote=FALSE,  sep="\t",  row.names=
FALSE, col.names=FALSE)
>library (GO.db)
Sink (file="GOinfoMF.txt", type="output")
>for (go in enriched.MF[]){
+ print (GOTERM[[go]])
+cat ("-------------------------------\n")
+ }
sink ()
```

2. *Cellular Components*

```
>GO.CC=goseq(pwf, "hg19", "geneSymbol", test.
cats=c("GO:CC"))
>enriched.CC=GO.CC$category[p.adjust(GO.CC$
over_represented_pvalue,method="BH")<.1]
>write.table(enriched.CC, file="GOenrichedCC.
txt", quote=FALSE, sep="\t", row.names=FALSE,
col.names=FALSE)
>library (GO.db)
sink (file="GOdescriptiveCC.txt", type="output")
> for (go in enriched.CC[]){
+ print (GOTERM[[go]])
+cat("---------------------------------\n")
+ }
sink ()
```

3. *Biological Processes*

```
>GO.BP=goseq(pwf, "hg19", "geneSymbol", test.
cats=c("GO:BP"))
>enriched.BP=GO.BP$category[p.adjust(GO.BP$
over_represented_pvalue,method="BH")<.1]
>write.table(enriched.BP, file="GOenrichedBP.txt",
quote=FALSE, sep="\t", row.names=FALSE, col.names=
FALSE)
>library (GO.db)
sink (file="GOinfoBP.txt", type="output")
>for (go in enriched.BP[]){
+ print (GOTERM[[go]])
+ cat("---------------------------------\n")
+ }
sink ()
q()
```

4   Notes

1. Users should be aware that there are a vast number of different tools and statistical tests available for the QC, alignment, read counting, and differential expression testing of RNA Seq data. Those used here are some of the most commonly used and publically available; however it is possible to incorporate different tools into parts of the workflow illustrated in the methods here (e.g., a different alignment tool than TopHat2).

2. The programs used in this chapter are continuously being updated and improved and therefore readers should ensure that they are working with the latest versions of each and check the accompanying manuals/vignettes for any changes that might occur.

3. The number of reads will also be calculated and displayed in the html output from FastQC.

4. It might be useful here to look at online examples of a good (http://www.bioinformatics.babraham.ac.uk/projects/fastqc/good_sequence_short_fastqc/fastqc_report.html) and bad (http://www.bioinformatics.babraham.ac.uk/projects/fastqc/bad_sequence_fastqc/fastqc_report.html) fastq dataset analyzed using FastQC.

5. If unsure about this some users might find it useful that FastQC attempts to automatically determine which encoding method was used to generate the quality scores for a given fastq file and this is displayed in the basic statistics section of the html file.

6. Users may wish to view their aligned sequences using an alignment visualization tool such as IGV (http://www.broadinstitute.org/software/igv/home). IGV can be started from a web browser with no need for installation. IGV requires BAM files to be sorted by position and indexed. The indexed .bam.bai files must be located within the same directory as the BAM file.

7. It is generally accepted that duplicate reads should be left in for the purposes of differential expression analysis of RNA Seq data; however this command might be useful for other purposes such as variant calling.

8. This is only necessary the first time you install the package unless you are performing an update.

9. It is also possible to import files from HTSeq count without creating a table of counts using the DESeqDataSetFromHTSeqCount function (see the vignette for further details).

10. Users should note that the default setting of DESeq2 performs some automatic functions such as independent filtering and detection of count outliers. Both of these can be turned off—see the vignette for details.

11. To perform enrichment analysis using KEGG pathways:

```
>KEGG=goseq(pwf,"hg19", "geneSymbol", test.cats=
"KEGG")
>enriched.kegg=KEGG$category[p.adjust(KEGG
$over_represented_pvalue,method="BH")<.1]
>write.table(enriched.kegg, file="KEGGenriched_.
txt", quote=FALSE, sep="\t", row.names=FALSE, col.
names=TRUE)
```

12. Users might also wish to separate this list into genes that have been identified as up- or downregulated and independently assess whether these genes are enriched for certain GO terms or pathways.

Acknowledgments

The authors acknowledge funding from Science Foundation Ireland grant 09/IN.1/B2640 to R.M.M.

References

1. Mortazavi A, Williams BA, McCue K, Schaeffer L, Wold B (2008) Mapping and quantifying mammalian transcriptomes by RNA-Seq. Nat Methods 5(7):621–628, http://www.nature.com/nmeth/journal/v5/n7/suppinfo/nmeth.1226_S1.html

2. Ozsolak F, Milos PM (2011) RNA sequencing: advances, challenges and opportunities. Nat Rev Genet 12(2):87–98

3. Wang Z, Gerstein M, Snyder M (2009) RNA-Seq: a revolutionary tool for transcriptomics. Nat Rev Genet 10(1):57–63

4. Smith RM, Webb A, Papp AC, Newman LC, Handelman SK, Suhy A, Mascarenhas R, Oberdick J, Sadee W (2013) Whole transcriptome RNA-Seq allelic expression in human brain. BMC Genomics 14:571. doi:10.1186/1471-2164-14-571

5. Heap GA, Yang JH, Downes K, Healy BC, Hunt KA, Bockett N, Franke L, Dubois PC, Mein CA, Dobson RJ, Albert TJ, Rodesch MJ, Clayton DG, Todd JA, van Heel DA, Plagnol V (2010) Genome-wide analysis of allelic expression imbalance in human primary cells by high-throughput transcriptome resequencing. Hum Mol Genet 19(1):122–134. doi:10.1093/hmg/ddp473

6. Sun W, Hu Y (2013) eQTL Mapping using RNA-seq data. Stat Biosci 5(1):198–219. doi:10.1007/s12561-012-9068-3

7. Park E, Williams B, Wold BJ, Mortazavi A (2012) RNA editing in the human ENCODE RNA-seq data. Genome Res 22(9):1626–1633. doi:10.1101/gr.134957.111

8. Zhao J, Ohsumi TK, Kung JT, Ogawa Y, Grau DJ, Sarma K, Song JJ, Kingston RE, Borowsky M, Lee JT (2010) Genome-wide identification of polycomb-associated RNAs by RIP-seq. Mol Cell 40(6):939–953. doi:10.1016/j.molcel.2010.12.011

9. Anders S, Reyes A, Huber W (2012) Detecting differential usage of exons from RNA-seq data. Genome Res 22(10):2008–2017. doi:10.1101/gr.133744.111

10. Trapnell C, Williams BA, Pertea G, Mortazavi A, Kwan G, van Baren MJ, Salzberg SL, Wold BJ, Pachter L (2010) Transcript assembly and quantification by RNA-Seq reveals unannotated transcripts and isoform switching during cell differentiation. Nat Biotechnol 28(5):511–515. doi:10.1038/nbt.1621

11. Vijay N, Poelstra JW, Künstner A, Wolf JBW (2013) Challenges and strategies in transcriptome assembly and differential gene expression quantification. A comprehensive in silico assessment of RNA-seq experiments. Mol Ecol 22(3):620–634. doi:10.1111/mec.12014

12. Wang Y, Ghaffari N, Johnson CD, Braga-Neto UM, Wang H, Chen R, Zhou H (2011) Evaluation of the coverage and depth of transcriptome by RNA-Seq in chickens. BMC Bioinformatics 10(12 Suppl):S5. doi:10.1186/1471-2105-12-s10-s5

13. Rapaport F, Khanin R, Liang Y, Krek A, Zumbo P, Mason CE, Socci ND, Betel D (2013) Comprehensive evaluation of differential expression analysis methods for RNA Seq data. Genome Biol 14(9):R95, http://arXivorg/abs/13015277v2

14. Busby MA, Stewart C, Miller CA, Grzeda KR, Marth GT (2013) Scotty: a web tool for designing RNA-Seq experiments to measure differential gene expression. Bioinformatics 29(5):656–657. doi:10.1093/bioinformatics/btt015

15. Hart SN, Therneau TM, Zhang Y, Poland GA, Kocher JP (2013) Calculating sample size estimates for RNA sequencing data. J Comput Biol 12:970–978. doi:10.1089/cmb.2012.0283

16. Wang L, Wang S, Li W (2012) RSeQC: quality control of RNA-seq experiments. Bioinformatics 28(16):2184–2185. doi:10.1093/bioinformatics/bts356

17. Morgan M, Anders S, Lawrence M, Aboyoun P, Pagès H, Gentleman R (2009) ShortRead: a bioconductor package for input, quality assessment and exploration of high-throughput sequence data. Bioinformatics 25(19):2607–2608. doi:10.1093/bioinformatics/btp450

18. Kim D, Pertea G, Trapnell C, Pimentel H, Kelley R, Salzberg SL (2013) TopHat2: accurate alignment of transcriptomes in the presence of insertions, deletions and gene fusions. Genome Biol 14(4):R36. doi:10.1186/gb-2013-14-4-r36

19. Trapnell C, Roberts A, Goff L, Pertea G, Kim D, Kelley DR, Pimentel H, Salzberg SL, Rinn JL, Pachter L (2012) Differential gene and transcript expression analysis of RNA-seq experiments with TopHat and Cufflinks. Nat Protoc 7(3):562–578. doi:10.1038/nprot.2012.016

20. Delhomme N, Padioleau I, Furlong EE, Steinmetz LM (2012) easyRNASeq: a bioconductor package for processing RNA-Seq data. Bioinformatics 28(19):2532–2533. doi:10.1093/bioinformatics/bts477

21. Dennis G Jr, Sherman BT, Hosack DA, Yang J, Gao W, Lane HC, Lempicki RA (2003) DAVID: Database for Annotation, Visualization, and Integrated Discovery. Genome Biol 4(5):P3

22. Väremo L, Nielsen J, Nookaew I (2013) Enriching the gene set analysis of genome-wide data by incorporating directionality of gene expression and combining statistical hypotheses and methods. Nucleic Acids Res. doi:10.1093/nar/gkt111

INDEX

Anthony W. Ryan (ed.), *Celiac Disease: Methods and Protocols*, Methods in Molecular Biology, vol. 1326,
DOI 10.1007/978-1-4939-2839-2, © Springer Science+Business Media New York 2015

Printed by Printforce, the Netherlands